Accomplishing Climate Governance

This book provides a new approach to thinking about the politics and geographies of climate governance. It argues that in order to understand the nature and potential of the range of new responses to climate change emerging at multiple scales, we need to examine how governance is accomplished – how it is undertaken, practised and contested. Through a range of case studies drawn from communities, corporations and local government, the book examines how climate change comes to be governed and made to matter as an issue with which diverse publics should be concerned. It concludes that rather than seeking the solution to climate change once and for all, we need to engage with the ways in which we can channel our intentions to ameliorate the climate problem to more progressive ends. The book will be of interest to researchers, advanced students and policy makers across the social sciences.

HARRIET BULKELEY is a professor in the Department of Geography, Durham University. Her research focuses on the processes and politics of environmental governance. Her recent books include *Governing Climate Change* (2010); *Cities and Low Carbon Transitions* (2011); *Cities and Climate Change* (2013); *Governing the Climate: New Approaches to Rationality, Power and Politics* (2014); *Transnational Climate Governance* (2014); and *An Urban Politics of Climate Change* (2015). She was awarded a Philip Leverhulme Prize in 2007 for her 'outstanding contribution' to geographical research and an Economic and Social Research Council (ESRC) Climate Change Leadership Fellowship in 2008 for her innovative work on cities and climate change. She has been an editor for the journal *Environment and Planning C* since 2008. She is currently involved in researching the politics and practice of smart grids in the United Kingdom, low carbon transitions in southern Africa, and continuing work on urban responses to climate change through the JPI Urban and ESRC project Governing Urban Sustainability Transitions. Harriet has undertaken commissioned research for the Joseph Rowntree Foundation, Friends of the Earth, UN-Habitat and the World Bank. In 2014, she was awarded the King Carl XVI Gustaf's Professorship in Environmental Science and a visiting professorship at Lund University, Sweden.

ACCOMPLISHING CLIMATE GOVERNANCE

Harriet Bulkeley
Durham University

CAMBRIDGE
UNIVERSITY PRESS

CAMBRIDGE
UNIVERSITY PRESS

32 Avenue of the Americas, New York, NY 10013-2473, USA

Cambridge University Press is part of the University of Cambridge.

It furthers the University's mission by disseminating knowledge in the pursuit of education, learning, and research at the highest international levels of excellence.

www.cambridge.org
Information on this title: www.cambridge.org/9781107038653

First published 2016

A catalog record for this publication is available from the British Library.

Library of Congress Cataloging in Publication Data
Bulkeley, Harriet, 1972–
Accomplishing climate governance / Harriet Bulkeley, Durham University.
pages cm
Includes bibliographical references and index.
ISBN 978-1-107-03865-3 (Hardback: alk. paper)
1. Climatic changes–Government policy–International cooperation. I. Title.
QC903.B49 2016
363.738′74561–dc23 2015014002

ISBN 978-1-107-03865-3 Hardback

For Jane and Rip Bulkeley
Who taught me that life is worth fighting for

> . . . as small and quiet and unimportant as our fighting may look, perhaps we might all work together . . . and break out of the prisons of our own making. Perhaps we might be able to keep this fierce and beautiful world of ours as free *for all of us* as it seemed to be on that blue afternoon of my childhood.

Cressida Cowell (2005: 238) *How to Speak Dragonese,*
Hodder Childrens Books, emphasis in the original

Contents

Acknowledgements

The idea for this book began, though I did not know it at the time, two decades ago. In the spring of 1995, I was finishing my undergraduate geography degree and hoping for funding to start my graduate studies in the autumn. Applying to the United Kingdom's Social Science Research Council for a project to examine Australia's climate politics across multiple scales, I argued for its relevance to the United Kingdom, where, whilst not so controversial, governing climate change was going to be challenging. They were not so convinced. A decade later, having secured funding for the Ph.D. project and subsequent work exploring how cities were responding to climate change, the question of the United Kingdom's climate governance reemerged. Now that we were beginning to make the argument that the governing of climate change was not confined to the international arena, what were we to make of it appearing everywhere? From celebrities to mayors, supermarkets to petrol chains, climate change appeared to be making the running. What did this mean? How could we understand it? And what were the implications for the social science of climate change?

It was with these questions in mind that I initially approached the Royal Geographical Society with the Institute of British Geographers with a small pilot project examining the transnational politics of climate change. In 2007, I was awarded a Philip Leverhulme Prize by the Leverhulme Trust, through which I continued to develop this work. I am enormously grateful to the Trust for its support, which enabled the empirical work and conceptual thinking to progress incrementally from 2008 to 2012 as I iteratively explored the terrain of the new politics and new geographies of climate change in the United Kingdom without the sense of an impending deadline and a defined list of necessary outputs. During this time, Andrea Armstrong (Durham University) and Sara Fuller (Macquarie

University) worked with me on the project. They conducted interviews, read material, wrote reports and provided a good deal of insight into the projects that they encountered. I would like to express my sincere thanks for their excellent work and good humour throughout this process. The book simply would not have been possible without their input. The last part of the project, writing the book, has been a similarly drawn-out process. It has taken place in between a number of other projects that have competed for time and attention. Matt Lloyd and Holly Turner at Cambridge University Press have been both encouraging and firmly patient during this time, and I would like to express my thanks to them for bearing with me. Over the final year of writing the book, I was partly based at Lund University, supported by the award of the seventeenth King Carl XVI Gustaf's Professorship in Environmental Science and a visiting professorship. This award allowed the final ideas for the book to develop and for the manuscript to come to fruition, and I would like to thank the Swedish Academy of Sciences and HRH King Carl XVI Gustaf for providing this wonderful opportunity.

Whilst this book is ostensibly the work of a single author, it is of course a truism that there is no such thing. Given the lengthy period over which the ideas developed here have gestated, it is perhaps little surprise that they have been provoked and influenced by my encounters with many good friends and colleagues. I would especially like to mention, at the Department of Geography, Durham University, Louise Amoore, Ben Anderson, Gavin Bridge, Nicky Gregson, Cheryl McEwan, Colin McFarlane, Joe Painter, Marcus Power and Simon Marvin, and further afield John Allen, Andrew Barry, JoAnn Carmin, Andrew Jordan, Diana Liverman, Susan Owens, Heike Schroeder, Matt Watson and the members of the Leverhulme International Network on Transnational Climate Change Governance for the time and energy they have spent discussing ideas with me over the years. During this period, I have worked closely with a number of collaborators who have given me much more support than I can ever acknowledge fully. The work here, though they might not recognise it as such, has been very much influenced by Michele Betsill, Vanesa Castan Broto, Robyn Dowling, Gareth Edwards, Matthew Hoffmann, Heather Lovell, Pauline McGuirk, Pete Newell, Matthew Paterson and Gareth Powells, and I would like to thank them for sharing their thoughts, critiques and friendship. I am also grateful to Heather Lovell and Johannes Stripple for reading parts of the manuscript and ensuring that

I was on the right tracks. It is not a coincidence that many of the formative stages of the work presented here took place in Lund. It was through engaging with Johannes' Ph.D. work that I first started to formulate my ideas around authorisation as practice. Through the COST Action on Transforming Global Environmental Governance, we were able to organise and assemble one of the first gatherings of social scientists concerned with developing more critical theoretical perspectives on questions of climate change at a workshop in Lund in 2011. The outline of this book was sketched one early June morning with the sun coming through the windows waiting for the workshop to start in the White House in the centre of Lund. Through the ongoing work on the collection that came from that workshop and our subsequent collaboration under the King Carl XVI Gustaf's professorship, Johannes has provided both the inspiration and the grounding without which this book would have been neither as developed in its argument nor as much fun to write.

Almost always last on the pages of acknowledgements, but never on a day-to-day basis, are my family. This book is dedicated to my parents, to whom my immeasurable debt becomes ever more apparent as both they and I grow older. At home in Durham, Pete, Elodie and Thea – who are now growing up rather used to the idea of a book in the house – live with ideas as my constant distraction whilst continuing to remind me of all of the more important distractions that life has to offer. Their love, patience and enduring interest, manifest in drawing my attention to the very small and the very big questions in life, enable me to accomplish everything I do.

I was on the right tracks. It is not a coincidence that many of the formative stages of the work presented here took place in Lund. It was through [illegible] Ph.D. work that I first started to formulate my ideas around authorisation as practice. Through the COST Action on [illegible] Global [illegible], we were able to organise [illegible] and assemble one of the first gatherings of [illegible] with [illegible] climate change at a workshop in [illegible] in 2011. The origins of this book were sketched out [illegible] June [illegible] with the sun [illegible] through the window [illegible]

1

Climate Problematics

Introduction

Climate change now needs little introduction. Through the production of scientific assessments, the gathering of international politicians, the iconography of environmental campaigns, commercial arenas from advertising hoardings to supermarket aisles, an impressive array of art forms, vernacular discourse surrounding everything from the weather to the price of gas and the insertion of new infrastructures across our urban and rural landscapes, climate change makes its presence felt. Climate change has gathered an ephemeral sense of ubiquity, highly (geographically and politically) uneven, dispersed yet somehow centralised, everywhere and at the same time difficult to locate. Its status as the wicked environmental problem sine qua non is seemingly unassailable, attracting sustained resources from across research, practitioner and policy communities, the attention of the world's media and (at least some of) the public. At the same time, a legion of potential solutions to the rising levels of atmospheric greenhouse gases and the impacts of the consequent changing climate have been proposed, from new international institutions, alternative financial allocations, innovative policy instruments, technologies, plans, resilient infrastructures, forms of behavioural change and so on and so forth. Indeed, most of the academic literature on the governing of climate change is concerned with such problem-solving interventions, their design, implementation, economies and politics.

Whilst much has been revealed about the nature and implications of climate change starting from this point, its diagnosis as an environmental problem and the concomitant responses offered are also problematic. Within the discipline of geography, questions of the environment have occupied a central if sometimes uneasy position. Despite the growing recognition since the wave of new environmentalism in the 1960s that matters of the environment are also acutely political, economic, social and cultural, 'the tendency to label some issues as "environmental" and others as "economic", "cultural" or whatever . . . dies hard' (Castree 2002: 358). The result, as Karen O'Brien argues, is that 'human geographers [and other social scientists] have failed to shift the focus of the scientific discourse away from "the environment" as the problem and towards an integrated understanding of [societal] change based on critical research on space, place, politics, power, culture identities, emotions, connections, and so on' (O'Brien 2012: 593–4). There are growing calls for an increased engagement of the social sciences in global environmental change research and policy processes designed to bring some traction to addressing climate change. Yet these largely remain committed to a paradigm in which society can be understood as responding to problems taking place in a separate biophysical world and where issues of power and inequality are subsumed by more practical considerations of how to design appropriate institutional, market or behavioural responses (Castree et al. 2014; Head & Gibson 2012; Swyngedouw 2010).

This book argues that to engage with climate change in alternative ways that acknowledge 'the inconsistencies and ambiguities that stalk the phrase' (Brace & Geoghegan 2010: 285), the multiplicity of meanings, forms and relations to which it gives rise, and its radical nature, requires that we start from a different place. Many different starting points are available, and work across the social sciences is beginning to explore these avenues and the new directions for thinking about climate change that they provide, from issues of security to those of emotion, the ethics of care and questions of daily practice. This book begins with one such entry point: the matter of climate change's politics. This is far from virgin territory, largely occupied by the disciplines of international relations and political science, with their concerns for the design of international institutions and the nature of national politics (Bulkeley & Newell 2015). Here, for the most part, climate change is treated as an object, a biophysical condition, to which various social entities – actors, institutions, policies – respond. Yet if we take the perspective that such a priori divisions between the social and natural are themselves subject to question, as is now commonly accepted across the discipline of geography and beyond, attention necessarily shifts to the ways in which climate change comes to be made political and how, in turn, political conditions are made in relation to climate change.

Central to these dynamics has been the matter of how climate change has come to be governed – of how climate governance is accomplished. Governing is not the only set of political conditions to which climate change gives rise and through which it comes to be constituted. There are, for example, forms of violence, security and conflict configured in relation to climate change. Yet the question of by what means and by whom climate change should be governed has preoccupied analysts and policy makers alike. Entering into this domain of inquiry, this book takes these questions in a different direction. Rather than considering governance as a matter of actors and institutions, it starts from the analysis of governing as the orchestration of distinct modes of power and seeks to explore the workings, politics and geographies of its operation (Allen 2003; Bulkeley 2012). Governing is itself a multivalent concept. Its familiar sense, associated with the organisation of modern democratic states, can be traced back to its roots as captured by Hobbes in *Leviathan*, where the term 'to govern' was articulated as the capacity 'to rule with authority, that is to rule with some basis' (Dean 2007: 36). Yet an alternative interpretation of governing can also be traced to classical political thought. As Dean (2007: 36) argues, a second, now more obscure, use of the term was also advanced by Hobbes, in *Behemoth*, as any 'practice that more or less deliberately seeks to direct, guide, or control others, for example, children, subjects, wives, a congregation, even live-stock, and so forth'. It is from this sense of governing that Foucault's formulation of government as the 'conduct of conduct' is also derived. Tracing multiple sources of the term, Foucault argues that before it acquired its particular political connotations in the sixteenth century governing had broad connotations. Indeed, he suggests that in the writing of the time:

> 'to govern' covers a very rich semantic domain in which it refers to movement in space, material subsistence, diet, the care given to an individual . . . and also to the exercise of command, of a constant, zealous, active and always benevolent prescriptive activity. It refers to the control one may exercise over oneself and others, over someone's body, soul and behavior. And finally it refers to an intercourse, to a circular process or process of exchange between one thing and another. . . . through all these meanings . . . one never governs a state, a territory or a political structure. Those whom one governs are people, individuals, or groups (Foucault 2009: 122).

Foucault's writings and associated works on governmentality provide a rich set of conceptual resources through which to examine what governing entails and, critically, how it is accomplished. Drawing on these perspectives, this book aims to move beyond an account of climate change as an issue to be governed and instead asks how it comes to be constituted as requiring intervention, how this is mobilized, sustained and contested. This introduction sets out the ways in which the governing of climate change has commonly been conceived and how a focus on governing as an accomplishment can bring to bear new insight. The chapter then turns to consider what such perspectives might entail for examining the politics and geographies of climate governance before introducing the remaining chapters of the book.

Governing the Globe?

As climate change has come to be a matter not only of scientific but of political concern, a growing academic literature has addressed the question of its governance. Largely derived from international relations accounts of the means by which states cooperate and how non-state actors are able to shape the governing of global affairs, a rich seam of research has revealed the multilevel, multi-actor and multi-issue nature of climate change governance. Yet such work often tends to bypass the more fundamental questions of why, how and with what effects governing takes place. Alternative perspectives, gathered from an eclectic set of traditions in social sciences that can broadly be labelled as critical political and social theory, because of their disposition to open up taken-for-granted concepts and positions, provide a different set of entry points based on an account of power that is able to examine the relational, material and contested nature of governing. From this perspective, understanding the nature of climate governance requires that we attend to how it is accomplished and contested across a diverse array of sites and arenas.

Towards Global Governance

The now commonplace framing of climate change as a global problem requiring collective solutions owes its roots to the emergence of the climate change agenda in the late 1980s as one requiring an international scientific and policy response (Paterson 1996). Conceived as one of several global environmental issues requiring international action, climate change came to be regarded as problem of the atmospheric commons that required new institutions capable of managing the challenge of ensuring effective action. During the 1990s, analysis of the politics of climate change was subsequently dominated by accounts of the formation of international regimes and the ways in which nation-states and other actors, from scientific communities to nongovernmental organisations (NGOs) and business interests, were able to effectively shape and direct global climate politics (Bulkeley & Newell 2015). Largely determined by the division within political science between studies of the international arena on the one hand, seen as the domain of international relations, and studies of domestic politics on the other, regarded as the arena of political science proper, accounts of the climate change regime as the locus of climate politics were soon subject to critique (Newell and Paterson 1998; Paterson 1996). Adopting the language of governance, which was at the time gaining popularity across the social sciences, research began to analyse climate change politics through a set of different frames that regarded the global nature of the issue not in absolute terms but rather in relation to the globalising and networked nature of the economy and society. Such analyses began to explore the ways in which other scales and arenas of decision making came to matter in relation to climate change, the roles of non-state actors not only in shaping the actions of states and their international institutions but in responding to climate change on their own terms and the ways in which the governing of climate change extended beyond the specific policy domain of the international agreement and related to different economic spheres (Bulkeley & Newell 2015; Okereke et al. 2009). Far from being confined to the international activities of the state, global governance could include 'a vast array of rule systems that exercise authority in the pursuit of goals and that function outside normal national jurisdictions' (Rosenau 2000: 172). Consequently, the global governance of climate change encompassed 'all purposeful mechanisms and measures aimed at steering social systems towards preventing, mitigating, or adapting to the risks posed by climate change' (Jagers & Stripple 2003: 385).

During the 2000s, a burgeoning literature emerged tracing and analysing the global governance of climate change through multiple domains, levels and arenas of decision making, actors and institutions (for reviews, see Biermann & Pattberg 2008; Bulkeley & Newell 2015; Newell et al.

2012). Yet freed from the confines of the international sphere and state-based institutions, what constituted global climate governance appeared rather hard to pin down, leading to protests that 'such all-encompassing definitions hardly leave room for anything that is not global governance' (Biermann & Pattberg 2008: 279). Rather than global climate governance being an amorphous concept, Biermann and Pattberg (2008: 279) suggest that it is the 'distinct *qualities* of current world politics, such as non-hierarchical steering modes and the inclusion of private actors, both for profit and nonprofit' (emphasis added), that should be taken to epitomise global (environmental) governance. As they go on to elaborate, global environmental governance is in this reading taken to be a new phenomenon characterised by three particular features:

> ... first, the emergence of new types of agency and of actors in addition to national governments, the traditional core actors in international environmental politics; second, the emergence of new mechanisms and institutions of global environmental governance that go beyond traditional forms of state-led, treaty-based regimes; and third, increasing segmentation and fragmentation of the overall governance system across levels and functional spheres (Biermann & Pattberg 2008: 280).

It is these features and their dynamics that lend governance a different set of qualities to its counterpart, usually taken to be traditional forms of state-based regulation or government. As Newell et al. (2012: 373) elaborate with regard to the role of non-state actors, it is not the presence of so many actors in the arenas of global environmental governance per se that matters but rather their ability 'to effectively steer particular aspects of the world political system in certain directions'. These features of global environmental governance – as multi-actor, multimodal and fragmented – are frequently read as emerging from the changing nature of the global economy and concomitant shifts in the nature of power and authority amongst different actors and institutions (Hoffmann 2011). The reach and legitimacy of the state are interpreted as waning in the face of the complexity of global governance challenges, creating the need to engage new actors and new modes of governing that in turn reflect 'shifts in the distribution of power and resources in the global political economy: trends toward the growing power of international nongovernmental organizations (NGOs) with transnational networks ... trends toward urbanization ... and the power of transnational capital in conditions of globalization' (Newell et al. 2012: 366).

These arguments, and others like them, have led to a growing interest in a host of forms of climate change governance that lie beyond the traditional remit of the state and international regimes, from transnational networks to cities; to the role of private authorities, including both corporate and environmental organisations; and to community-based organisations (Bulkeley & Newell 2015). For some, however, this interest in the 'new' climate governance has gone too far, for in the 'rush to study new forms of governing "beyond", "below" and "outside" the state-dominated climate regime', the critical role of the state in providing a source of governance innovation in this domain is being neglected (Jordan & Hutimea 2014: 388; see also Biermann & Dingwerth 2004). Certainly, much of the research on climate governance can appear to be concentrated on arenas in which the state has only a shadowy presence. Yet this framing of climate governance as being either of one kind or another, of the state or non-state, in itself needs further critical examination. Both the 'old' and the 'new' debates on climate governance have a tendency to regard the state – and indeed other actors, institutions, scales, spheres of the social world and so on – as predetermined and discrete entities. Climate governance is here the sum of the parts – all of the mechanisms and measures being used to steer societies in response to climate change (Jagers & Stripple 2003). In this manner, climate change governance effectively describes a set of actors, institutions, arrangements, interventions and instruments that together – because of their distinct qualities – can be taken to represent the governance of a particular domain that we have come to know as climate change. Turf wars may ensue within this field as to the proper interpretation of the role of different actors (state vs. non-state), authority (public vs. private) and scale of decision making (global vs. national vs. local), but on the whole, climate change governance is understood as pertaining to these largely social responses to a set of biophysical or material conditions of the changing composition and functioning of the climate system.

From this perspective, the tendency amongst analysts has been to ask things *of* climate governance. How effective might one or other arrangement be? What are the means by which innovations can be spread? How legitimate are the actors involved? In what ways can the transparency of decision-making processes be improved? These are indeed laudable questions and have come to be more prominent as social scientists have been tasked with becoming more involved in the design and implementation of solutions to the climate problem. Yet in focusing analysis on these dimensions of governance, questions *about* governance have often been strangely neglected. For example, much of the debate, on both sides of the state/non-state divide, tends to operate with a conception of the state as a reified 'thing: a more or less unified entity' (Painter 2006: 754) which then 'interacts with', 'intervenes in', 'depends upon' or 'regulates' other distinct social spheres such as 'the economy'[or] 'civil society' (Painter 2006: 753). Shifts in the nature of authority and its fragmentation have largely then been interpreted in zero-sum terms as if the gain of one actor is a loss to another (Bulkeley & Schroeder 2012; Sending & Neumann 2006), without further interrogation as to how the state and non-state might be conceived differently. Likewise, when it comes to the shifting geographies and fragmented politics of climate change governance, 'for all this talk of a redistribution or shift in capabilities between the different levels of governance ... the vocabulary of power is still one of capabilities "held" and the dispersion or distribution of powers between various levels and sites of authority' (Allen 2004: 22). A critical engagement with climate change governance requires that, rather than taking the nature of the entities and arrangements through which it is conducted for granted, these are opened to scrutiny. This is not only a matter of asking how, why, by and for whom climate change is governed, but of interrogating how what it means to govern is constituted in relation to climate change.

Accomplishing Governance

The starting point for a more critical engagement with the governance of climate change has been to regard governance not as a set of actors and institutions, but in terms of the specific modes of power through which governing is conducted and the processes and practices through which this takes place. Early work in the climate change field in this vein came from the neo-Gramscian tradition of political economy and its concern with the intimate relations amongst the state, private actors and civil society (Levy & Egan 1998; Levy & Newell 2005; Newell & Paterson 1998). Rather than regarding the state as a bounded entity, neatly cleaved from other spheres of the social world, such a perspective entails a notion of the state as a system of 'strategic selectivity' (Jessop 2002) in which it is composed of 'all institutions which enable the dominant social groups to exercise power, whether formally public or private, as components of the state's institutional ensemble' (McGuirk 2004: 1022). Such perspectives have been central to developing our understanding of the ways in which climate change has come to be governed and the political economies of the workings of, for example, carbon markets, offset certification and corporate social responsibility. Yet whilst the boundaries between the state/non-state and public/private are troubled by the critical political economy perspective, there remains a sense that these are definite areas of the social world whose boundaries can be readily charted. Moreover, the power to govern is concentrated in the hands of elites, operating as a resource or capacity for some to wield over others and universalised so as to be of the same nature whatever the matter at hand.

There are, however, other ways of conceiving of the nature and workings of power that open up the questions of its mobilisation, politics and geographies in different ways. Whilst it is often taken for granted that power is 'an instrument of domination, a capacity of some resourceful mix' (Allen 2010: 2899), alternative approaches stress instead that power is facilitative, 'generated by the application of resources and skills over tracts of space and time' (Allen 2010: 2900; see also Barnett & Duvall 2005; Lipschutz 2005), and immanent, 'inseparable from its effects' (Allen 2003: 65). Moreover, as has long been recognised in political science, 'the currency of power is not the same for all relations' (Hurd 1999: 379). In his sustained engagement with the geographies of power, John Allen (2003, 2004, 2010) has shown that power takes different modalities, from domination and coercion, to authority, seduction, inducement and so on, in turn requiring that we attend to the 'distinctive characteristics and circumscribed consequences' (Allen 2003: 117) of its orchestration. What such accounts suggest is that the distinct qualities of governance as a social (and material) relation come not from the actors and institutions, or indeed structures, involved but from the particular modalities of power at work. Governance is not all-encompassing – some relations are of a different nature – but its distinctiveness is to be found in the ways in which power, conceived here in facilitative and immanent terms, is orchestrated and contested.

Foucault's concept of governmentality provides an important means through which to conceive of the orchestration, processes and practices of governing (Miller & Rose 1990; Stripple & Bulkeley 2014). Regarded as one mode of power, operating in the modern state alongside sovereignty and discipline (Foucault 2009: 105), the art of government is understood as the 'the conduct of conduct', 'modes of action, more or less considered and calculated, that were destined to act upon the possibilities of action of other people' (Foucault 2000a: 341). Rather than seeking to secure particular territories or control individuals, government seeks to 'improve the condition of the population, to increase its wealth, longevity and its health' (Foucault, 2009: 105). This 'will to improve the condition of the population, is expansive' and tends to encompass multiple spheres of social life (Li 2007a: 5–6). The 'will to improve' (Li 2007a) can take multiple forms in different domains of the population but is achieved through orchestrating, or conducting, the everyday actions, or conduct, of individuals. Governmentality, in turn, is regarded as 'the ensemble formed by institutions, procedures, analyses and reflections, calculations and tactics that allow the exercise of this very specific, albeit very complex, power' (Foucault 2009: 108).

The notion of governmentality has been subject to multiple interpretations across the social sciences. Partly because of its fragmented treatment in Foucault's own work but also because of the nature of his approach, any definitive interpretation is rather elusive. For William Walters (2012: 45), rather than simply applying the concept, what is required is a critical encounter with the set of openings provided by the notion of governmentality: 'rather than a set of arguments to be endlessly rehashed or interpreted, hidden meanings or references to be uncovered, we can relate to it as something to be used, adapted, set to work in grappling with problems'. For the purposes of understanding the processes and practices of governing advanced in this book, three such encounters are especially productive. First, governing is not seen as a set of discourses or techniques that are applied to a particular problem area, but rather the field of intervention – what it is that should be governed and how this should be done – is itself constituted through the process of governing. The working of government as a mode of power entails constituting a particular will to improve in relation amongst broad political rationalities, such as neoliberalism; the problematisation of particular conditions of the population; and a set of practices, techniques or technologies of government that are 'not just governed by institutions, prescribed by ideologies' but also 'up to a point, possess their own specific regularities, logic, strategy, self-evidence and "reason"' (Foucault 2000b: 225). The ensemble of these rationalities and techniques, and the particular governmentality they form, has variously been termed an apparatus (Walters 2012), assemblage or dispositif (Li 2007b), and forms a heterogeneous network of elements and devices through which the problem to be addressed and the kinds of intervention required are constituted. Under conditions of liberal and neoliberal government, governmental assemblages tend to expand, accreting new elements 'pulled together from an existing repertoire, a matter of habit, accretion, and bricolage' (Li 2007a: 6). It is through the configuration, orchestration and contestation of the dispositif/assemblage/apparatus that governmentality operates, establishing the 'right way of arranging (disposer) things in order to lead (conduire) ... not to the form of the "common good" ... but to a "suitable end", an end suitable for each of the things to be governed' (Foucault 2009: 99). Governing is then not a matter of implementing solutions to predefined problems, but rather of the constitution and configuration of what should be governed and what it means to govern in tandem.

If this first encounter with governmentality serves to emphasise how delimiting what constitutes the problem requiring intervention is itself central to the work of governing and how it is an unfolding, heterogeneous and practical endeavour, a second encounter brings us to its sociomaterial nature (Barry 2001, 2013; Bulkeley et al. 2007). The sociomaterial character of governing can be found in at least two aspects of the concept of governmentality. For one, Foucault's argument that governing is concerned with the population is not restricted to actors and their institutions, but rather extends to 'a sort of complex of men and things' (Foucault 2009: 96), including 'wealth, resources, means of subsistence, and of course, the territory with its borders, qualities, climate, dryness, fertility ... customs, habits, ways of acting and thinking ... accidents, misfortunes, famine, epidemics and so on' (Foucault 2009: 96). This is not simply a matter of expanding the repertoire of things that need to be governed, as if they were an amalgam, but rather of attending to their iterative relations. Famously drawing on the metaphor of the ship, he explains: 'what characterizes government of a ship is the practice of establishing relations between the sailors, the vessel, ... the cargo ... and their relations with all those eventualities like winds, reefs, storms and so on' (Foucault 2009: 97). The intricate interrelation between men and things is also exemplified in Foucault's example of the governing of the town, in which he argues that it forms a milieu, 'a field of intervention in which, instead of affecting individuals as a set of legal subjects capable of voluntary actions ... one tries to affect, precisely, a population ... a multiplicity of individuals who are and fundamentally and essentially only exist biologically bound to the materiality within which they live' (Foucault 2009: 21; see also Bulkeley et al. 2015).

Beyond regarding the objects of governing in sociomaterial terms, Foucault's work also points to the critical sociomaterial quality of the work of governing. Governing is not only ideological or discursive, but takes place through 'apparently humble and mundane mechanisms which appear to make it possible to govern' (Miller & Rose 1990: 8). Many of the mechanisms, techniques or technologies of government that have been subject to scrutiny by scholars of governmentality, such as calculation, audit, accounting, participation and so forth, have not been considered in material terms. Yet in keeping with a wider trend across the social sciences, the sociomaterial constitution and potential of such mechanisms are increasingly recognised as central to the ways in which governing is accomplished. As Barry (2001: 5) argues, 'in Foucault's account, government is inevitably a technical matter. Practices of government rely on an array of more or less formalised and more or less specialised technical devices from car seat-belts and driving codes to dietary regimes; and from economic instruments to psychotherapy.'

A third encounter with governmentality that provides insight into how the governing of climate change is accomplished relates to the way in which the constitution of subjects/subjectivities is central to the conduct of conduct (Paterson & Stripple 2010). Rather than being determined a priori, like the fields of intervention that constitute the objects of governance, subjects are constituted through the processes and practices of governing. Subjectivities, the condition of being (a) subject, its qualities and disposition, are constituted through the articulation of the rationalities and techniques of governing and their sociomaterial manifestation discussed previously, and in turn serve to shape and delimit the nature of intervention and what it is that improving the condition of the population might entail. This in turn means that conduct is not orchestrated by force, coercion or regulation, for 'it is not a matter of imposing a law on men, but of the *disposition of things* . . . arranging things so that this or that end may be achieved through a certain number of means' (Foucault 2009: 99, emphasis added). As Li (2007a: 5) elaborates, at the level of the population, 'it is not possible to coerce every individual and regulate their actions in minute detail. Rather government operates by educating desires and configuring habits, aspirations and beliefs.' Recent assessments suggest that there has been relatively limited engagement with this aspect of Foucault's work within human geography and across the broader social sciences (Paterson & Stripple 2010; Rutherford 2007). Where such approaches have been adopted, for the most part, it has been concerned with the ways in which different instances of neoliberal governmentality, characterised by various calculative devices, have served to create particular subjectivities of individuals and communities as responsible for their own welfare, well-being and improvement.

However, 'the dominant tendency to focus on governmental rationalities' in such analyses 'contributes to a view of power as top down, totalizing and omnipresent' to the neglect of the messy actualities of governmental practice (McKee 2011: 3) and the ways in which different kinds of subjectivity are constituted. As 'geographers' engagements with governmentality have become increasingly complex', there is an 'emerging scepticism towards the coherence of neoliberalism as a programme of rule and the inevitability of fully formed neoliberal subjects' (Dowling 2010: 493), which makes space for considering alternative subject positions and the workings of the productive forms of power and its resistance in Foucault's original formulation of the notion of conduct (Rutherford 2007). Rather than being subject to coercion or discipline, the practice of governmentality relies on the condition of freedom 'in which several kinds of conduct' are possible (Foucault 2000a: 342). Moreover, once power is conceived in relational, productive terms and as orchestrated through conduct, conduct in turn becomes a site for 'resistance, refusal, or revolt': a suite of forms of more or less explicit ways of contesting and avoiding particular intentions through what Foucault comes to term 'counter-conduct in the sense of struggle against the processes implemented for conducting others' (Foucault 2009: 200–1).

Rather than being a 'secure accomplishment' (Li 2007a: 10), the certain outcome of the actions of particular actors and mechanisms, these encounters with the concept of governmentality and associated forms of critical social science suggest that governance is more productively conceived as a programme, an open-ended and thoroughly contested set of processes and practices through which governing is pursued. Governing is programmatic, not only because it is accompanied by a 'proliferation of more or less explicit programmes for reforming reality' but also because it is accompanied by 'an eternal optimism that a particular domain or society could be administered better or more effectively, that reality is, in some way or other, programmable' (Miller & Rose 1990: 4). Rather than attending to the features of the climate change governance landscape and their suitability or otherwise for administering particular climate solutions, a critical engagement requires that we examine the dynamics through which governing is produced – to the ways in which climate governance is accomplished. To accomplish entails both to 'fulfil, perform . . . achieve' a set of desires or objectives and to 'complete (a journey)' in one sense and to gather and perfect particular skills in another (*Oxford English Dictionary* 2015). Understanding how governance is accomplished is then a matter of examining how it is performed, fulfilled and completed in relation to different desires and objectives through the acquisition and honing of particular skills and techniques. The conceptual repertoire of governmentality provides some important starting points for such an analysis, but, as recent scholarship has illuminated, it is a set of tools rarely used in isolation (Stripple & Bulkeley 2014). In offering a new set of directions for work on the governing of climate change that start from a concern with the ways in which power, in its facilitative and immanent sense, is orchestrated, practised and resisted, governmentality offers a new set of entry points into the debate without determining the direction of inquiry. In the remainder of this book, these encounters are used to explore both why and how climate change has come to be considered as requiring interventions of all sorts, and to examine the kinds of politics and geographies through which intervention is constituted and in turn serves to configure.

New Politics, New Geographies?

The politics and geographies of the existing order of climate change governance have been extensively charted. Where power is conceived as a held and finite capacity, politics is both an outcome – who gets what, where and how – and central to the zero-sum games through which actors seek to wrestle power from one another. In climate change terms, it has been the politics of the international negotiations that have attracted the most attention. Whilst the literature on global governance has done much to widen the scope of actors considered to be relevant to these politics, at the heart of the discussion remains the question of which states are able to command sufficient resources to sway the regime in one or another direction. Recent discussions on the climate complex have pointed to the multiple arenas through which this politics takes place but have left the question of climate's politics largely untouched (Abbott 2012; Keohane & Victor 2011). Climate change is in these readings a classically and fiercely political issue. However, for others nothing could be further from the truth. Far from being an overtly contested domain, climate change is the ultimate post-political issue (Kenis & Mathijs 2014; Swyngedouw 2010). In this line of argument, the emergence of climate change is associated with a (neoliberal) consensus about its apocalyptic consequences and the need for collective action in response as part of a wider condition that has served to evacuate 'dispute and disagreement from the spaces of public encounter to be replaced by a consensually established frame . . . structured around the perceived inevitability of capitalism and a market economy as the basic organizational structure of the social and economic order, for which there is no alternative' (Swyngedouw 2010: 216). Climate change is post-political in as much as its governance foregoes the possibility of dissent and the potential of (radical) alternatives to the status quo of political and economic orders.

Whilst there are clearly fundamental differences in the articulation of what constitutes the politics of climate change between these perspectives, they are in some senses more similar than they at first appear. For in each case what is political about climate change is the way in which the power struggles take place in more or less public fora and what is required to ensure the proper conduct of politics is to be achieved through a distinct public or democratic realm. For those concerned with politics 'red in tooth and claw' in the international negotiations, more democratic forms of decision making are required in the form of transparency, accountability and new mechanisms for engaging those who have to date been excluded. Despite the contrast in their position, those concerned with climate change as post-political also rely on democracy as the means through which to resuscitate climate politics, though here 'a recuperated democracy' is seen 'as the terrain (space) for expressing conflict, for nurturing agonistic debate and disagreement, and, most importantly, for the naming of different possible socio-environmental futures' (Swyngedouw 2010: 229). Both perspectives bring important insight into the political nature of climate change, yet in limiting their conception of power to either the resources of particular actors or the imprint of structural conditions, so too what serves as the space for the political is constricted.

This constriction of what constitutes political space is perhaps particularly ironic at a time when analyses of the geographies of climate's governance have become more sophisticated. Dominated during the 1990s by the iconography of the globe, the geographies of governing of climate change were initially confined to the international realm. Inasmuch as they featured at all, other geographical scales were seen in hierarchical terms such that policies and mechanisms agreed upon at the global scale would cascade down to national, regional and local arenas to be implemented (or ignored) (Bulkeley 2005). Seeking to challenge this dominant paradigm, the global governance debate, discussed previously, sought to widen the range and scope of these geographies and to employ the notion of multilevel governance to point to the interwoven scales and networks through which the governing of climate change was taking place (Betsill & Bulkeley 2006; Gustavsson et al. 2009). Yet despite the growing range of sites and arenas considered as engaging the governing of climate change, for the most part these remain rather fixed territories, delimited in advance of the operation of climate governance above or through their domain. Here too the continued emphasis on an approach to governance that privileges a conception of power as a fixed and divisible set of capacities reinforces a particular sense of its geographies as a matter of the 'spatial extent or scale over which it may be exercised' (Bulkeley 2012: 2433). This sense of power as extendable over space (and time) is 'a more or less ingrained part of our geographical imaginations. Ingrained, in the sense that we are accustomed to thinking about power as, on the whole, an observable feature of any given territory or that its extension across a flat surface is, for the most part, relatively unproblematic' (Allen 2011: 285). Whilst the emergence of new forms of global climate governance shifts the boundaries of how power is allocated and extended, it has so far done little to change the terms of debate and open up the possibilities that the geographies of climate change governance could be otherwise.

Shifting the terms of climate change governance requires then only that we engage not only with new conceptions of its process and practice – with how governing is accomplished – but also with alternative means through which to understand its politics and geographies. Starting from the encounters with the field of governmentality undertaken earlier in this chapter, this book argues for a threefold reconceptualization of the politics of climate change governance. First, if power is conceived not as a held resource but in facilitative and immanent terms, its politics emerges not in terms of how it is shared or in the turf wars about its ownership, but rather in terms of how its productive and constitutive force is orchestrated. Rather than being divisible in any sense, power is mutually and relationally constituted and as such its politics are also those of mutuality and its tensions, a 'permanent provocation' to the means through which governing is pursued (Foucault 2000a: 342; Li 2007a: 11). Rather than existing a priori, power is generated through the new forms of relation constituted in the governing of climate change so that at stake in the practice of such assemblages is how and by what means such relations are configured and sustained. This suggests that any such configuration is far from stable but in a continual process of being closed and opened, 'punctured by a challenge it cannot contain' (Li 2007a: 11), reassembled and renewed. This in turn means that what may appear to hold as consensus or as conflict must be subject to scrutiny and the dynamic ways in which such precarious achievements come to be stabilised and interrogated.

Second, the emphasis on governing as a programmatic, open-ended and technical process suggests that rather than think of politics as a reified or romanticised notion, we need to attend to its practicalities, to adopt 'a view of politics as a craft or skill involving "power, its acquisition, distribution, and use" (Geuss 2008, 96)' (Agnew 2011: 470). Politics, in this sense, is not confined to distinct arenas or appropriated by some kinds of actors (public) over others (private). Rather, politics can be found in the workings of power wherever this might be situated, whether this be in the mundane, such as the routine practices of traffic management in the urban arena, or in the spectacular, from the corridors of climate negotiations to the protest camps of anticapitalist movements. Third, if governing is a sociomaterial condition, so too are its politics. As Barry (2001: 5) argues, 'the study of government [in the Foucauldian sense] . . . opens up a much broader field of politics to inspection. The political need not be only associated with the control of political institutions, the activities of the state or the formation of social movements. Instead . . . the political [may] refer to the ways in which artefacts, activities or practices become objects of contestation.' Such forms of contestation may be highly visible and at the same time, as Barry (2013) finds in terms of the geopolitics of the development and installation of transnational oil pipelines, form around quite specific technical matters, the visibility of which is itself subject to dispute. At the same time, making things more visible – or bringing controversy into the open – can also serve as a strategy for depoliticisation under the guises of technical apparatuses of transparency and accountability, for example (Barry 2013). There is then no straightforward relation between open antagonism and political space, but rather what constitutes the political has to be forged through the processes and practice of governing wherever its whereabouts.

In seeking to open up the terms of inquiry for a new politics of climate change governance, questions of its geographies are also bought into view. Renewing our understandings of the geographies of climate change governance requires more than acknowledging their multiplicity; it requires that we attend to the ways in which particular orderings of the world are created, sustained and contested in relation to climate change and how in turn climate change is constituted through sociospatial ordering. From this perspective, this book seeks to make two contributions to how we conceive and understand the geographies of climate change governance. First, it seeks to extend the argument that power is not merely projected over or through space but rather that '*spatiality is constitutive of power relations not only in general, but also in the particular ways in which different modes of power take effect*' (Allen 2003: 102, emphasis in the original). Rather than being fixed in territorial or other geographical terms, 'particular spatialities are constituted and (re)produced through the mediated socio-technical relations within which power is effectuated' (Bulkeley 2012: 2433). The qualities of particular forms of sociospatial relation are therefore not given, but rather a 'laborious work in progress, prone to failure and permeated by tension and contradiction. ... never complete, but always becoming' (Painter 2010: 1094). This in turn means that the analytical task is not one of tracing the geographical boundaries of particular forms of climate governance arrangement, or even of seeking to understand how they might be mobilised across space or between scales, but rather of seeking to understand why, how and to what effect distinct spatial orderings (individual, community, shopping centre, transnational network) configure and are configured by the will to improve climate change.

Second, this book attends to the ways in which such spatial orderings are constituted topologically through the practices of bricolage, assemblage and contestation central to how governing is accomplished. Such a reading of the spatiality of government is central to the way in which Foucault distinguishes discipline and the apparatuses of security or governmental power:

> Discipline is essentially centripetal . . . [it] functions to the extent that it isolates a space, that it determines a segment. Discipline concentrates, focuses and encloses. . . . the apparatuses of security . . . have a constant tendency to expand; they are centrifugal. New elements are constantly being integrated. . . . Security therefore involves organizing, or anyway allowing the development of ever-wider circuits (Foucault 2009: 44–5).

Rather than government being conducted at a distance, as if it were exercised uniformly over space, such an account points to the ways in which governmental programmes extend through circulations that require and in turn foster the folding together of disparate elements, drawn from near and far into various configurations of reach and proximity (Allen 2011; Allen & Cochrane 2010). Yet such configurations and circulations are not random, blindly gathering all before them, but riven through the particular design and practice of governmental programmes and the rationalities and techniques that they deploy. For example, as Mitchell (2008: 1118) explains, it was an 'apparatus of calculation that bought elements from other projects – generator costs, transmission losses, data on gas lighting, population densities – into Edison's workshops at Menlo Park, New Jersey, where it could be combined with other instruments of measurement and calculation in the construction of an electrical system', which were then in turn relayed beyond this particular site and moment to the workings of electricity systems in the wider world. Tracing the new geographies of climate change governance requires working through the ways in which entities are bought into proximity or alignment and animated through particular sociospatial relations, whether that be as communities, multinational corporations, cities, local places or so on, and how this composition in turn shapes the ways in which the politics of governing climate change are conducted.

Outline

In seeking to develop these arguments, this book situates its investigation of how climate change governance is accomplished in the context of the United Kingdom. Since climate change first came to be regarded as a political, economic and social issue as well as a scientific one, the United Kingdom has featured as a prominent actor in the international negotiations and as an arena within which policy innovation has taken place. Whilst the definitive account of the United Kingdom government's policy response to climate change across these different scales has yet to be written, this is not the project attempted here. Rather, the book takes as its empirical focus a set of six loosely connected interventions that were all, in one way or another, designed to act on and for climate change. These range from the actions of multinational corporations, to government programmes and community-based activities. Chapter 2 provides an overview of how, why and by what means climate change has come to be a concern across these different arenas, charting the development of different agendas and programmes over the past three decades. At the same time, it examines how state-based actors, corporations and communities have in turn become central to the way in which climate change has become regarded as a problem and its solution sought. Through this analysis, the chapter argues that climate governance can be characterised by a 'paradox of slow progress' (Owens 2010), where a fortuitous historical reduction in greenhouse gas emissions, ambitious intentions and radical change are coupled with contradictions, fragmentation and precariousness. Across this patchwork landscape, the chapter finds that the intention to govern climate change is itself fraught, pulled in different directions and subject to ongoing contestation and reorientation. This intention fails to be contained within particular channels or programmes, but instead is subject to ever-increasing circulation and extension. The chapter closes by introducing the six cases explored in the book, from the banking, energy and retail corporate sectors, and from rural and urban communities.

Chapter 3 takes up one of four analytical themes deployed to examine the processes and practices of accomplishing climate governance: authorisation. Authorisation is frequently taken for granted as central to the workings of governance. To govern, indeed, is to 'seek authority for one's authority' (Rose 1999, quoted in Li 2007a: 12). Yet the nature of authority is rarely full examined. Analysis tends to be concerned with who holds authority, its extent, limits and contestation. Where authority is seen as one form of power that can be shared, analysts have tended to ask how authority is shifting between different actors – be it the rising powers of the new economies of China, India, Brazil and so on, or forms of private authority working within or across state boundaries. Working with the conceptual repertoire developed in this first chapter, Chapter 3 takes a different starting point and seeks to understand how and to what effect authorisation is produced and practised. It argues that authorisation is practised through distinct modalities 'constituted through the purposes for which power is deployed, the nature of recognition and compliance, and distinct forms of socio-spatial relations' (Bulkeley 2012: 2434). Developing three such modalities – of authorisation as consent, consensus and concord – the chapter explores the dominance of consensus in the cases of governing of climate change examined in this book. Rather than signalling the emergence of a generalizable post-political condition, these cases point to the ways in which resistance and contestation are coincident with the working of consensus, creating new political terrains.

Authorisation is critical to the ways in which governing is accomplished, but it is insufficient. Governing requires an impetus or intervention, a problem requiring resolution and a will to improve (Li 2007a). Particular sociomaterial conditions have to be found wanting or containing some form of promise. These conditions are not simply those of the real world, but rather have to be made in relation to climate change. It is to this matter of bringing order to climate change that Chapter 4 turns. It suggests that the shift from the broad will or intent to govern climate change to specific programmes of intervention is neither straightforward nor guaranteed but requires ongoing work to keep the climate problem coherent and legible, and that this will be subject to continual forms of reconfiguration, reversal and contestation. Investigating this dynamic, the chapter examines two related processes. First, it explores how interventions are assembled through a suite of practices that both serve to problematize climate change and to align it with and/or make it through particular organisational forms and cultures. It argues that the making of governmental programmes is not only a calculated activity but also one of aspiration and desire. Second, it develops these arguments by considering the ways in which climate change comes to be comprehended, where to comprehend involves not only knowing climate change, but also how it is caught, contained and embraced. This analysis reveals that what is involved in making climate change comprehendible are not only those kinds of practices that render it intelligible – subject to rational or intellectual thought – but also the means through which it becomes sensible – felt in an emotional register.

Chapter 5 turns more fully to analyse the practical and technical nature of governing climate change, of the ways in which it comes to be articulated through the disposition of 'men and things'. Whilst analysts have pointed to the array of techniques and devices through which disparate elements come to be aligned, configured and related across space and time, it has been those concerned with calculation in its various guises and the ways in which forms of commensuration orchestrate carbon markets that have attracted the most attention. The analysis in this book finds that such forms of calculation are not only quantitative, but also pertain to the qualities that are realised through working on climate change, seeking not only to operate and mobilise governing at a distance but through the proximities and contingences of particular sociomaterial conjunctures. Likewise, commensuration, 'making things the same', has been achieved through techniques, practices and devices that enact and enable a 'logic of equivalence' (Li 2011) such that new entities and values mobilised through governmental programmes are entrained within existing configurations. These forms of governmental work come under pressure on the one hand as the specificity of climate change responses needs to be retained, certified and exemplified, and on the other as equivalence is questioned and matters of incommensurability overflow the intervention.

The analysis turns, in Chapter 6, to examine the ways in which the governing of climate change is a public matter. Conventionally, the argument that climate change is of public concern has been derived from its taken-for-granted status as a common goods problem requiring collective management. In seeking to shift the terrain of this debate, Paterson's (2013) analysis draws on the work of Hannah Arendt to argue that what constitutes the climate (re)public is made through practices of deliberation and intersubjective learning. Rather than being given, climate's public but has to be made, fashioned through the ways in which governing is accomplished. Yet rather than focusing on processes of political deliberation, largely outside of the dominion of private life and primarily orchestrated by forms of discursive exchange, this chapter considers the ways in which different sociomaterial entities are related as matters of climate concern as 'objects, devices, settings and materials, not just subjects, acquire explicit political capacities, capacities that are themselves the object of public struggle and contestation, and serve to enact distinctive ideals of citizenship and participation' (Marres & Lezaun 2011: 490). Assembling climate publics, it is argued, is an avowedly sociomaterial accomplishment. In Chapter 7, the concluding chapter of the book, the key arguments concerning the ways in which governing is accomplished through authorisation, ordering, articulation and making public are revisited and their political and geographical implications considered. Rather than seeking grand designs for climate changed futures, it argues that in order to further accomplish climate governance, we need to attend to the 'small and quiet and unimportant' ways in which it is being configured and contested in order to open up their progressive potential.

2

Charting Climate Change Governance in the United Kingdom

Introduction

Since climate change first made its way on to political and public agendas in the late 1980s, the United Kingdom has attracted considerable attention for its international and domestic action. The United Kingdom's scientific communities were central to the formation of the Intergovernmental Panel on Climate Change in 1988 and the establishment of the Hadley Centre, through which extensive global climate modelling has since taken place. Margaret Thatcher, then prime minister, attracted widespread attention for her speeches to the Royal Society in 1988 and the UN General Assembly in 1989 warning of the long-term and fundamental changes to the environment taking place and calling for collective action to resolve global warming. The United Kingdom government subsequently played a key role in negotiating the text for the 1992 United Nations Framework Convention on Climate Change and the 1997 Kyoto Protocol. As part of the European Union, the United Kingdom government has led the call for ambitious international targets and timetables to reduce greenhouse gas emissions in the subsequent decades and continues to be at the forefront of international climate diplomacy. This international position has been enabled through the United Kingdom's fortuitous position domestically, where the privatisation of the energy industry and a political desire to shift away from a dependence on coal under the Conservative government of the 1980s led to what has been termed the 'dash for gas' for power generation and the subsequent decline in greenhouse gas emissions during the early 1990s (Collier 1997; Eyre 2001; O'Riordan & Rowbotham 1996). Whilst these developments marked promising beginnings for an active response to climate change, throughout the 1990s and early 2000s commentators raised concerns about the lack of concerted effort to tackle climate change domestically and the growing level of greenhouse gas emissions. Yet by the mid-2000s, climate change was once again at the heart of the political agenda as it became inextricably linked with the future of the United Kingdom's energy sector and increasingly its economic development through the growing rhetoric of the need to foster a transition to a low carbon economy (Lovell et al. 2009). In 2008, after sustained pressure to address the United Kingdom's growing emissions from the private sector and environmental groups, this intention was crystallised in the Climate Change Act, which introduced the legally binding obligation on the United Kingdom government to reduce greenhouse emissions to 80 per cent below 1990 levels by 2050.

Internationally and domestically, the United Kingdom's climate change policy responses have appeared both ambitious and central to maintaining momentum for a global climate agreement. Yet at the same time there has been much criticism of the absence of strong policy responses within the United Kingdom. This has been levelled at the contradictory nature of policies that have also been introduced to expand road building, develop land for housing and encourage increasing levels of consumption, as well as those that have undermined efforts to develop renewable energy resources, and the failure to tackle energy demand. Climate policy appears fragmented, fragile and readily overturned by more mainstream concerns. This has been particularly apparent in the past five years as the Conservative–Liberal Democrat coalition government has struggled with the implications of implementing the Climate Change Act and climate sceptics have gained more of a political voice. This concomitant sense of both radical change and limited advance is what Susan Owens (2010: 394) terms a 'paradox of slow progress' where 'at any one moment, progressive change in environmental governance can seem painfully slow, sometimes even unattainable. But when we look back over periods of a decade or more, we can see that the policy and political landscapes have changed beyond all recognition.' Climate governance in the United Kingdom epitomises this paradox, not only in terms of the international and domestic responses adopted by the central government but more broadly in terms of how climate change has come to be understood and acted upon across a whole host of sites and arenas from corporate boardrooms to village halls.

This paradox of ambitious intentions and radical change coupled with contradictions, fragmentation and fragility provides a fertile ground through which to examine how climate governance is accomplished. What is evident is that there is no straightforward association between political intention and governing realities. Moreover, the intention to govern climate change is itself fraught, pulled in different directions and subject to ongoing contestation and reorientation. Nor does the governing of climate change contain itself in either the channels of international diplomacy or the arena of national policy making, but rather spills out across such boundaries. New and often unlikely conjunctions of moments of opportunity, political entrepreneurs, policy issues, climatic emblems and arenas are drawn together in order that climate governance finds expression. The rest of this chapter sets out some of the key moments, issues and arenas through which climate change has come to be framed, understood and acted upon in the United Kingdom in order to provide the contextual background through which the analysis of accomplishing climate governance in a series of cases can be read. The first section relates the story of the development of national climate change policy in the United Kingdom, whilst the second section turns to the role of the private sector and the third to the role of local and community responses. In each case, three historical periods are apparent: the late 1980s until 1997, a period that witnessed the emergence of the climate change agenda and the negotiation of the Kyoto Protocol internationally and the policies of the ruling Conservative government in the United Kingdom; from 1997 to 2005, the period over which the Kyoto Protocol entered into force and the new Labour government initiated a series of policy measures in the United Kingdom whilst private and community sector responses began to develop; and the period after 2006, described by Carter and Jacobs (2014) as involving radical policy change, which has since been followed by significant contestation. The final section of the chapter introduces the case studies that form the analytical core of the subsequent chapters and sets them in this broader context.

The Shifting Terrain of National Climate Policy in the United Kingdom: From Environment to Energy and Economy

It was in 1988 that the confluence of growing scientific evidence, extreme weather events, the statement by the chief climate scientist at NASA, James Hansen, to the U.S. Senate that climate change demanded action, a conference in Toronto 'On the Changing Atmosphere', Margaret Thatcher's speech at the Royal Society and the formation of the Intergovernmental

Panel on Climate Change (IPCC) conspired to move climate change onto the political agenda (Hulme & Turnpenny 2004; Paterson 1996). Several countries (and cities) began to adopt unilateral targets to reduce greenhouse gas emissions. The United Kingdom's 1990 sustainable development strategy, *This Common Inheritance* (Department of the Environment 1990), included the first government commitment towards responding to climate change with a target of reducing carbon dioxide to 1990 levels by 2005. Whilst climate change was firmly framed as an environmental issue, associated with other global environmental problems such as ozone depletion and deforestation and with policy led by the Department of Environment, parallel developments in the energy sector were critical in shaping the United Kingdom's response. Under the Conservative government in the 1980s, many previously nationalised industries were privatised, including the rail and bus services, telecommunications, water, gas and, in 1989, electricity. The privatisation of the electricity industry saw a decoupling of production and supply, and the 'dash for gas' that led to a new round of investment in combined cycle gas power plants. Driven by cost efficiency and energy security, a crucial side effect has been to shift the United Kingdom's power system onto less fossil fuel–intensive forms of power generation. By the mid-2000s, 'overall UK [greenhouse gas] emissions fell by 21% between 1990 and 2007', with 'over 75% of this ... due to a decline in heavy industry and changes in fuel used for energy supply, from coal to gas' (Marsden & Rye 2010: 669), changes that largely took place in the early 1990s.

More specific efforts to address climate change were at this time driven by the process of international negotiations sparked by the growing scientific evidence about the nature of climate change and the momentum building around the 1992 Rio Earth Summit. Negotiations amongst the European Economic Community (the precursor to the European Union) and the drive to secure an intergovernmental agreement at Rio led the United Kingdom to adopt the more stringent goal of reducing emissions of carbon dioxide to 1990 levels by 2000. Following the successful negotiation and signature of the United Nations Framework Convention on Climate Change (UNFCCC), the United Kingdom government published its first Climate Change Programme (Department of the Environment 1994) – a collection of measures to be undertaken in order to achieve the goal of emissions reductions by 2000 (Lovell et al. 2009). At the heart of the approach was a focus on improving domestic energy efficiency, through the introduction of a value-added tax on domestic fuel and establishing the Energy Savings Trust as an independent government body to provide advice and information to the public on how they might save energy at home. However, these measures were restricted in scope and ambition and with 'emissions of greenhouse gases falling relative to 1990 levels, there was limited appetite for engaging in additional policies and measures' (Lovell et al. 2009: 95).

The coincidence of the international negotiations for the Kyoto Protocol and the election of the Labour government in 1997 gave the United Kingdom's climate policy a degree of renewed impetus. Coming into office, Labour pledged to go beyond the United Kingdom's Kyoto target of reducing emissions to 12.5 per cent below 1990 levels by the period 2008–2012, with its own target of a 20 per cent reduction by 2010 (Carter 2008; Hulme & Turnpenny 2004). At the same time, climate change began to move beyond the confines of the environmental policy arena through, for example, the creation of the Department for Environment, Transport and the Regions (headed by Labour Deputy Prime Minister John Prescott); the introduction of policies intended to shift away from a 'predict and provide' approach to the delivery of transport and housing that were partially couched in the discourse of needing to address rising greenhouse gas emissions; new directions for waste policy intended to move away from landfill and increase recycling; and experiments with the introduction of new policy instruments intended to increase the costs of carbon (Bulkeley & Betsill 2003). With the exception of the waste sector, this impetus to seek to reduce demand as a way of managing infrastructure, land and environmental issues came under significant pressure (Owens & Cowell 2011), and was quickly disbanded along with the Department for Environment, Transport and the Regions, which in 2001 was separated into the Department for Environment, Food, and Rural Affairs and the Department for Transport, Local Government and the Regions.

The Labour government released a new Climate Change Programme in 2000, containing both the promise of a new suite of practical instruments and acknowledgement of the long-term challenges ahead. These innovative policy instruments faired a little better than efforts to integrate climate change into other sectors. Although when then-chancellor Gordon Brown 'proposed the new Climate Change Levy in 1999 he was subjected to an extremely hostile, well-organized business lobby demanding that his proposals be watered down' (Carter 2014: 424), a version of this policy instrument was nonetheless introduced in 2001. Designed to target the largest energy users, the Climate Change Levy was a tax on business energy use offset by a reduction in their contribution to National Insurance payments. The most energy-intensive business sectors negotiated an alternative Climate Change Agreement, whereby they were exempt from 80 per cent of the Climate Change Levy subject to attaining agreed emissions reductions commitments (Lorenzoni et al. 2008: 106–7). Accompanying these measures, a pilot Emissions Trading Scheme was established in 2002 in the United Kingdom with thirty-four large energy users agreeing to cap their emissions voluntarily and undertake various measures to meet reduction targets or trade emissions with one another. At the same time, in 2002 a Renewables Obligation was launched to oblige suppliers to provide electricity from renewable sources, and the Energy Efficiency Commitment/Carbon Emissions Reduction Target was established to charge energy providers with improving levels of energy efficiency in the housing stock. Despite these notable forms of innovation, climate change policy in the early 2000s remained an isolated matter of (largely) environmental concern, with attempts to engage traditional policy sectors, including transport, housing and energy, with the climate issue largely rebuffed.

This ground began to shift with the publication in 2000 of *Energy in a Changing Climate,* an influential report by the Royal Commission on Environmental Pollution, which urged the United Kingdom to adopt a long-term and radical approach to decarbonisation. Established in 1970 (and abolished in 2011 under the Coalition government), the Royal Commission on Environmental Pollution (RCEP) was an advisory body of experts charged with providing independent review of critical environmental issues for government. It was in 1997 that the RCEP turned its attention to the topic of energy and environment, driven by the growing scientific evidence surrounding the issue of climate change; a sense that wealth economies such as the United Kingdom would need to bear the burden of significant reductions in greenhouse gas emissions; and also the imperative of 'a discursive strategy which framed the issues in a compelling way, so as not to produce just "one more report on energy"' (Owens 2010: 398). Shaped by these drivers, at the heart of *Energy in a Changing Climate* was the notion of undertaking a fundamental transformation of the United Kingdom's energy system such that the use of fossil fuels was radically reduced to 60 per cent below 2000 levels by 2050.

As Susan Owens (2010: 396), herself a member of the RCEP at the time, argues 'to the Commission's own surprise, the Government picked up the gauntlet – and did so swiftly in comparison with previous occasions on which reports had significantly challenged the status quo' by charging the Performance and Innovation Unit (a strategy unit for the Cabinet Office, at the heart of government) with developing policy objectives for the long-term future of energy policy in the United Kingdom. Published in 2002, *The Energy Review* (PIU 2002) signalled for the first time that addressing climate change would require fundamental and long-term changes in the United Kingdom's energy system and acknowledged the critical role of the RCEP in shaping the direction of government policy (Lovell et al. 2009; Owens 2010).

Following the publication of *The Energy Review*, in 2003 the government released the first Energy White Paper since 1967, which placed climate change squarely at the heart of energy policy whilst also signalling the critical role of the energy system in achieving climate policy goals. The White Paper, *Our Energy Future: Creating a Low Carbon Economy* (DTI 2003), contained a commitment by the government to meet the RCEP's goal of reducing the use of fossil fuels in the energy system by 60 per cent below 2000 levels by 2050 and focused on increasing renewable energy and improving energy efficiency. Critics, however, strongly argued that rather than resolving these policy challenges, the Energy White Paper lacked detailed or new policies and measures (Lovell et al. 2009). At the same time, rising energy prices and concerns about the security of gas supplies coupled with growing evidence that the United Kingdom would potentially struggle to meet the target of reducing emissions by 20 per cent below 1990 levels by 2000 set by the Labour government served to ensure that rather than settling the debate, the Energy White Paper effectively opened up a new series of debates about the nature of the United Kingdom's energy and climate policy.

The subsequent period, particularly between 2006 and 2010, has been heralded as one of 'radical policy change' for the United Kingdom's climate policy (Carter & Jacobs 2014). Whilst taking the long view it is possible to see that such shifts emerge gradually over time and are always incomplete (Owens 2010), it is certainly the case that the mid-2000s was marked by a growing intensity of events and policies concerned with the United Kingdom's response to climate change. Despite the growing evidence of increasing levels of greenhouse gas (GHG) emissions within the United Kingdom, Prime Minister Tony Blair made climate change the centre of the G8 meeting hosted in Gleneagles in 2005, seeking to continue to position the United Kingdom as an international leader in this policy arena. A revised Climate Change Programme was published in 2006 and, acknowledging that emissions would likely decrease only by 15–18 per cent below 1990 levels by 2010, sought to reinvigorate government commitment to the range of measures and instruments that had been introduced in the earlier period of the Labour government whilst also endorsing the 2005 European-wide Emissions Trading Scheme and introducing some important new measures, such as funding for piloting Carbon Capture and Storage, a new Code for Sustainable Homes, and funding for microgeneration technologies. Concerned by the lack of progress and subject to increasing critique in terms of the policy approach that was being adopted, the government also committed to a new Energy Review in 2005, just two years after the publication of the 2003 Energy White Paper. Controversially, the resulting report, *The Energy Challenge* (DTI 2006), opened up the possibility of new nuclear power generation whilst also continuing to support the need for Carbon Capture and Storage and the production of large-scale renewables (Lorenzoni et al. 2008). The report emphasised the need to develop more distributed forms of energy generation and to promote energy efficiency, as well as improving the efficiency of vehicles and appliances. For the most part, therefore, the review continued to emphasise the kind of responses that had characterised the Labour government's approach to climate change since it came to office and continued to attract criticism for its lack of ambition and foresight (Carter & Jacobs 2014).

That this position of encouraging gradual change in the energy, industrial, transport and housing sectors should shift to one in which the government committed itself, through the Climate Change Act of 2008, to meeting legally binding targets for emissions reductions towards a goal of reducing emissions by 80 per cent below 2000 levels by 2050, was the result of the conjunction of multiple climate events from both inside and outside government. Commissioned by the then-chancellor Gordon Brown, the *Stern Review* released in 2006 has been credited with making the economic case for climate action and drawing increasing levels of support from across the private sector for a coherent policy approach (Stern 2007). Public opinion was also moving towards a more proactive approach, with the release in 2006 of Al Gore's *An Inconvenient Truth* and growing levels of media coverage around the issue (Carter 2014). Perhaps most importantly, in 2005 Friends of the Earth launched the Big Ask campaign with the aim of attracting cross-party support for a Climate Change Bill that would contain annual targets for the United Kingdom to reduce emissions by 3 per cent and catalysed the formation of the Stop Climate Chaos campaign, which attracted the support of more than one hundred organisations and groups, from Greenpeace to the Women's Institute (Carter 2014; FoE 2014). By autumn 2006, the campaign had attracted the support of more than four hundred Members of Parliament, including the opposition leader David Cameron, and the government announced its intention to publish a Climate Change Bill. Critical to this development was the appointment in May 2006 of David Miliband as secretary of state for the Department of the Environment, Food and Rural Affairs. Widely regarded at the time as a potential leader of the Labour Party and politically ambitious, Miliband presided over the development of the new Office for Climate Change, regarded as providing a neutral space for the development of policy across the Cabinet, and took up the issue of the Climate Change Bill within the government (Carter & Jacobs 2014; Rutter et al. 2012).

If 2006 was the 'year the world woke up ... to the vast economic, political and social implications of climate change' (*Guardian* 2006: 7), it was a year later that the United Kingdom government's new response began to crystallise in the form of the 2007 Energy White Paper and the Draft Climate Bill. Whilst the 2007 Energy White Paper, *Meeting the Energy Challenge* (DTI 2007), continued to place strong emphasis on issues of security and affordability, climate change remained central to this policy domain, with renewed emphasis on the need for low carbon forms of energy supply, including nuclear power, and energy efficiency. In parallel, the development of the Climate Change Bill entailed negotiations over the nature of the targets and the process through which government should be held to account. The Climate Change Committee was formed as an independent body to provide advice to this process and in the autumn of 2008 recommended that the government adopt the target of an 80 per cent reduction in all United Kingdom emissions, including aviation and shipping, below 1990 levels by 2050, a target that was enshrined in the Climate Change Bill when it passed into law in October that year. The statutory targets within the Climate Change Bill were accompanied by the establishment of the Climate Change Committee, an independent advisory body that oversees government compliance with five yearly carbon budgets (Carter 2014). Shortly after the Climate Change Bill became legally binding, in December 2008, the United Kingdom also committed itself to the EU Energy and Climate Change Policy package, which set targets for 2020 of 'a 20% reduction in European CO2 emissions, a 20% increase in energy efficiency and 20% of final energy to come from renewable sources' (Foxon et al. 2010: 1209). For the United Kingdom, this entailed adopting a 'target of 15% of all energy to come from renewable sources by 2020 (compared to 2% in 2007), including around 30% of electricity' (Carter 2014: 423–4).

In the wake of these significant developments, the government shifted its approach to climate change policy away from publishing a programme of loosely related activities to a more focused *Low Carbon Transition Plan* (HM Government 2009) intended to deliver the '18% emission reductions [below 2008 levels] needed between 2009 and 2020' to meet the carbon budget of a reduction in over 30 per cent of emissions below 1990 levels through a range of measures, including 'renewable energy, feed-in tariffs, zero carbon homes, carbon capture and storage … infrastructure planning and domestic energy efficiency, and supported by significant public investment' (Carter 2014: 424). Rather than simply being a matter of enhancing the development of carbon markets, the *Low Carbon Transition Plan* recognises the critical roles of investment, infrastructure and innovation, as well as the roles of multiple actors, from the private sector to communities, in achieving these ambitions. Moreover, as the economic recession became apparent during the late 2000s, the low carbon economy became a means through which hopes for stimulating new forms of market and employment were vested, and in 'July 2009 the Government published a "low carbon industrial strategy", providing £405 million to support job creation and growth in "green" sectors' (Carter & Jacobs 2014: 135).

The incoming Conservative-Liberal coalition government – elected in 2010, having lent support in opposition to the Climate Change Bill and with Prime Minister David Cameron having declared that by 'voting blue', the United Kingdom could 'go green' – was already publicly committed to pursuing ambitious climate change policies. In 2011, the new government released *The Carbon Plan: Delivering Our Low Carbon Future* (HM Government 2011), which set out the means through which it intended to maintain progress towards the carbon reduction target for

2020 and to meet the target agreed for the fourth carbon budget period of 2023–2027 of reducing emissions to 50 per cent below 1990 levels (with a caveat to revisit these in 2014). *The Carbon Plan* advocates a mixed approach to achieving low carbon transitions, rather than picking winners in the form of particular technologies, but focuses on energy efficiency coupled with the electrification of heat and transport as well as the reconfiguration of electricity distribution networks to cope with this additional load. However, despite this positive engagement with the issues, climate and energy policy have come under intense pressure during the period since the election of the Coalition government. The Conservative right wing has openly contested the science of climate change and emphasised the potential economic implications of holding the United Kingdom to meeting the carbon budgets, particularly in the wake of the lack of internationally coordinated action following the failure to reach an international agreement at the Copenhagen Climate Summit. Growing opposition from the political classes to onshore wind farms has been accompanied by strong support for the development of the United Kingdom's fledgling shale gas fracking industry. At the same time, concerns for energy security have grown with the closure of some of the United Kingdom's nuclear and coal-fired power stations. Nonetheless, a number of measures have been introduced even in the face of this opposition, from the (ill-fated) Green Deal to provide financing for energy efficiency retrofitting, to the establishment of a Green Investment Bank, the mandating of the roll-out of smart meters to all United Kingdom households and support for the development of smart grids and decentralised energy generation through finance schemes such as the Renewable Heat Incentive. More significantly, in July 2014 the review of the fourth carbon budget concluded that the target should remain despite significant concerns voiced by Chancellor George Osborne (*Guardian* 2014a). By October 2014, the United Kingdom had signed up to the European Union's target of reducing GHG emissions by 40 per cent by 2030 and to the voluntary aims of achieving 27 per cent of energy generated from renewables and a 27 per cent improvement in energy efficiency (*Guardian* 2014b).

At the level of national government, climate change has come to occupy a central place on the political agenda, despite the waxing and waning of its fortunes. It is perhaps unsurprising that an issue that goes to the heart of the ways in which economies are organised has attracted significant political contestation. In the mid-2000s, commentators from research and policy communities were concerned that in this context the United Kingdom would struggle to meet its initial targets for reducing GHG emissions (Lorenzoni et al. 2008; Lovell et al. 2009). Yet the assessment of the Committee on Climate Change is that 'emissions in 2013 were 12% lower than 2007 and 28% below their 1990 level' (CCC 2014: 10). However, this progress should be treated with some caution. First, the Committee suggests that the pace of progress is not sufficient to meet longer-term targets and that further action is required to strengthen policies to go beyond the existing 2020 horizon that dominates policy thinking (CCC 2014: 10–1). Second, to date action by the UK government to intervene significantly and transform climate and energy policy has been forthcoming only when it has been feared that emissions levels are rising. It remains to be seen whether such action can be galvanised whilst emissions fall, albeit that such reductions are not adequate for meeting long-term targets. Finally, it is clear that concerns for energy security and affordability currently dominate the debate in this sector and that climate change remains relatively marginal in other critical policy arenas, including transportation, housing and economic development. Whilst climate change retains a higher profile than in many other nation-states, it is clear that the United Kingdom no longer commands a position of international leadership on either the European or the world stage, and its domestic policies are struggling to keep up with its political ambitions and legal commitments (Carter 2014; Lorenzoni et al. 2008).

Making Climate Change a Corporate Concern

The 'unprecedented politicization of climate change' (Carter & Jacobs 2014: 132) that took place in the United Kingdom in the mid-2000s was not confined to the arena of national government policy but was part of a widespread shift in the way in which climate change came to be related across multiple social and economic arenas. Central to the changing landscape of climate change governance was the shifting nature of the corporate response to the issue, and nowhere was this more critical than in the United Kingdom.

During the 1990s, the framing of climate change as largely a matter of scientific evidence, environmental concern and international negotiations meant that it was an issue largely ignored by the majority of corporate interests. However, two critical exceptions can be identified that served to play a significant role in shaping the evolution of climate policy. First, fossil fuel and related resource industries took a prominent stance in the international arena opposing action that would jeopardise their corporate interests (Bulkeley 2001; Newell & Paterson 1998). At the international level, these interests formed organised coalitions, notably the Global Climate Coalition and the Climate Council, the former of which included 'over fifty-five business trade associations and companies including the American Petroleum Institute, Du Pont, Dow, Ford, General Motors, Texaco, Chevron, Mobil and Shell' (Newell & Paterson 1998: 683). These coalitions had a significant influence over the development of international climate policy and, in particular, on the positions of key national governments, including the United States and Australia (Bulkeley 2001; Newell & Paterson 1998). The cohesion and influence of this grouping began to wane in the international sphere in 1997 as negotiations towards the Kyoto Protocol gathered pace. Famously, the then–group chief executive of BP, John Browne, broke ranks from the Global Climate Coalition in 1997, making a speech at Stanford University in which he supported the scientific consensus of the IPCC's second assessment report and advocated a partnership between businesses and government as a means of responding. BP's departure was followed by that of Dupont, Ford and Shell so that by the late 1990s the coalition increasingly lacked credibility and by 2002 it had disbanded.

The second set of corporate interests that came to play a significant role in the early period of climate policy response were the banking and insurance sectors. The antecedents to this can be traced to the 1991, 'when a small group of commercial banks, including Deutsche Bank, HSBC Holdings, Natwest, Royal Bank of Canada, and Westpac, joined forces with UNEP [UN Environmental Programme] to catalyse the banking industry's awareness of the environmental agenda' (UNEP 2014). In the run up to the 1992 Rio Earth Summit, these organisations formed the Banking Initiative, to promote awareness of environmental issues and foster interest in investment in the environmental domain. Following this, in 1995, lobbied heavily by then-director of Greenpeace, Jeremy Leggett (Carlsson et al. 2005: 256), 'UNEP joined forces with a group of leading insurance and reinsurance companies, including General Accident ... National Provident ... Swiss Re, as well as pension funds, to launch the UNEP Statement of Environmental Commitment by the Insurance

Industry. In this voluntary commitment, signatory companies pledge that they will aim at achieving a balance of economic development, the welfare of people and a sound environment' (UNEP 2014). Whilst framed in general terms around sustainable development, 'it is clear that global warming was driving the interests of insurers in such an initiative. By May 2000, the statement had 84 signatories from 27 countries' (Paterson 2001: 23).

At the heart of the concerns of the banking and insurance industries was the sense that climate change posed a series of risks to which they were financially exposed and that posed fundamental challenges to the nature of their industries. In addition, the environmental lobby, notably Greenpeace and UNEP, saw in these sectors a means of countering the power and influence of the fossil fuel industries in the eyes of national governments and as a means of securing investment in the new energy economy (Carlsson et al. 2005; Paterson 2001). One leading example was the work of Jeremy Leggett, then director of Greenpeace, in seeking to persuade the insurance industry to invest in solar technologies. In 'December 1996 Leggett organized the "Solar Investment Summit" in Oxford, designed to bring together insurers, bankers, and people from the solar energy industry, to stimulate investment in the latter' (Paterson 2001: 24). The organisation, Solar Century, established to provide a vehicle for this investment, failed to secure this level of interest and has since become a commercial enterprise, though it continues to lobby for investment in renewables. Whilst this initial attempt did not secure the level of investment in renewable energy that had been envisaged, it was an important first step in opening up the potential forms of corporate response to the climate change agenda.

Whilst clearly critical in lobbying governments and in shaping the nature of the climate change policy debate, what is striking about the nature of both the response from the fossil fuel and finance industries in the early 1990s is the way in which they were largely conducted separately from the development of national climate policy. Climate governance remained largely the province of national government and international negotiations, whilst corporate responses took place either within their own firm as a matter of corporate responsibility or through the formation of small (transnational) clubs or networks that sought to develop common responses and shared knowledge (Bulkeley et al. 2014). In the wake of the agreement of the Kyoto Protocol, however, this began to shift rapidly as the number of climate change governance experiments operating between state and local governments, across the public and private boundaries and transnationally rapidly increased (Bulkeley et al. 2014; Hoffmann 2011). Such initiatives were partly driven by the growth of carbon markets and their promotion, regulation and management, such as the activities of the World Business Council on Sustainable Development and the multiple forms of commercial organisations set up to account for and trade in carbon. Whilst such activities shaped the development of corporate responses to climate change in the United Kingdom during the early 2000s, through for example the activities of Eco-Securities and Climate Care, it was the development of a more general business engagement with climate change that proved critical in shaping the governance landscape.

Whilst some United Kingdom corporations, including Lloyds Insurance, HSBC, and BP, had been at the forefront of proactive responses to climate change during the 1990s, the sense within the corporate sector that climate change was an issue with which it ought to be concerned only gathered significant momentum in the early part of the 2000s. As Susan Owens (2010: 399) argues, throughout the 1990s and 2000s in the United Kingdom, 'diffuse, longer-term trends associated with ecological modernisation, including the shift towards precautionary, proactive approaches in environmental policy and the growing belief that green technologies could help reconcile economic growth and ecological integrity', were critical in opening up government policy making around energy and climate change. Such diffuse trends were also apparent in the corporate sector, as the notion that responding to climate change could be good for business caught hold. Capturing the essence of these arguments, then–Prime Minister Tony Blair gave a speech in London to members of the Prince of Wales's Business and Sustainability Programme (BSP), an initiative established in 1994 by the Prince of Wales and the University of Cambridge (Prince of Wales 2014). Following the speech, and Blair's call for a more engaged approach from the corporate sector, the Prince of Wales's Corporate Leaders Group on Climate Change was formed through this network. By its own account, 'its activities and achievements range from coming forward in support of a strong EU ETS [Emissions Trading System] and bold targets in an open letter to the Prime Minister in 2006, to developing and delivering The Copenhagen Communiqué on Climate Change in 2009, which saw over 950 global companies from 60 countries endorse the statement developed by the group, calling for an ambitious, legally binding and robust deal at COP-15' (Prince of Wales 2014).

Commentators have also pointed to the significance of the Corporate Leaders Group and its proactive approach in addressing the government through open letters in the press calling for more and sustained action on climate change. In particular, despite the group calling for a more active role for corporations, the actions of the group seem 'to have influenced Blair, who was genuinely surprised by their public letter in 2006 calling on the government to provide a clear and ambitious framework for the "transition to a low carbon economy"'(Carter 2008: 200). Another driver of the growing corporate response was the formation of the Climate Group in 2005, headquartered in London, and its focus on developing the economic case for addressing climate change (see Chapters 4 and 5). The *Stern Review* and the growing endorsement of the need for a clear response to climate change in 2006–2007 from a range of actors on the international stage, from politicians to movie stars, combined with the emerging carbon markets and the potential of the low carbon economy, served to shift climate change from an environmental concern to a matter of business for many corporations. For example, the effect in the retail sector was a shift that saw 'a much greater emphasis on absolute emission reduction targets and ... an explicit broadening of the focus of action from operational emissions to wider supply chain and value chain related emissions' (Gouldson & Sullivan 2012: 4). This was led by:

> Marks and Spencer (which committed to making its UK and Republic of Ireland operations carbon neutral by 2012 and to working with its customers and suppliers to help reduce their emissions) ... and Tesco (which committed to reducing its own carbon footprint and to working with its suppliers and other organisations to deliver significant greenhouse gas emission reductions across the supply chain ...). Five other retailers (Asda, Cooperative, Morrisons, Sainsbury's and Waitrose) followed suit, with each setting major/ambitious climate change targets in subsequent years (Gouldson & Sullivan 2012: 4).

Despite these prominent initiatives, not all corporate sectors have embraced the climate change agenda, and many in the energy-intensive sectors of the economy have actively resisted the development of climate policy over the past two decades. Whilst there has been active support for the development of industry initiatives such as emissions trading, as Hale (2010) points out, 'the call in 2006 by the Corporate Leaders Group for an ambitious cap in phase two of the emissions trading scheme ... did not contain a single company covered by the scheme.' This sector is critical to achievement of the United Kingdom's climate targets in the near and medium term, given that 'the UK has around 1,000 EU ETS participants' and the 'sectors covered by the EU ETS, will account for over 50% of the emissions reductions needed to meet UK targets between 2013 and 2020' (DECC 2014a). During the late 1990s and early 2000s, the Confederation of British Industry (CBI), the major representative organisation of British industry, remained opposed to the development of economic instruments that targeted carbon emissions reductions. Yet by 2007, even this long-standing concern had begun to wane as the CBI signalled its support for action on climate change, and this position has been entrenched such that the Confederation now suggests that climate change 'is not a threat to be feared, but an opportunity to grow the economy and lead the world' (CBI 2014). Large energy-intensive companies based in the United Kingdom, such as Unilever and Akzo Nobel, have joined with Forum for the Future, a not-for-profit organisation promoting sustainable development, in the Big Shift campaign to seek to move their business systems towards lower carbon futures (Forum for the Future 2014), suggesting that even in these hard-to-reach sectors of the economy, climate change and the low carbon transition have come to have an influence over the nature of what good business entails.

Since the initial responses of the United Kingdom corporate sector for and against international action on climate change in the early 1990s, the 'paradox of slow progress' identified by Owens (2010) is visible whereby what climate change means for the agenda of corporations in the United Kingdom has shifted from being an issue largely outside their remit to one with which they are often directly engaged either through the requirements of legislation or through their own volition. Accompanying this dynamic has been a change in the way in which corporations have sought to govern climate change, as many have moved from regarding climate change as largely the business of government and taken climate change to be central to the governing of business. Since the early part of the 2000s, a growing range of partnerships and initiatives have developed that have seen public–private collaboration and the development of private sector schemes intended to act on climate change, either through working with nongovernmental organisations, like-minded corporations or as innovations within individual corporations. This activity is beginning to have a profound effect on the landscape of climate change responses across the United Kingdom. For example, in the two decades since Jeremy Leggett organised the solar century conference in Oxford, the role of corporations in investing in renewables has shifted significantly from a position of disinterest to one of active involvement. In 2014, British Telecom, for instance, furthered its intention to use 100 per cent renewable electricity through signing power purchase agreements worth £440M with three large wind farms in the United Kingdom and investing in their further development. The provision of secure and low carbon energy is becoming a matter then not only for government but also for the corporate sector. Such developments have inevitably attracted accusations that corporations are undertaking such schemes as forms of greenwash, with critiques deploring a lack of ambition and the underlying imperative of profit and economic growth as undermining efforts to address climate change. Yet proponents argue that the pursuit of profit and the low carbon economy need not be opposed and indeed are necessarily linked (Climate Group 2014a). What is clear is that by the second decade of the twenty-first century, climate change has become firmly embedded as an issue for the body corporate, opening up the range of sites and arenas through which the governing of climate change is being accomplished.

Climate Change from the Bottom Up? Local and Community Responses in the United Kingdom

When climate change first came on to the agenda of the United Kingdom government, its framing as a global issue served to channel responses towards the creation of global scientific assessments and international political agreements. Yet, perhaps rather curiously, this framing also provided the basis for seeking the engagement of the United Kingdom's citizens through the launch in 1991 of a public information and advisory campaign, Helping the Earth Begins at Home, by the Department of the Environment (Hinchliffe 1996). The campaign drew on the rhetorical trope of the 1987 Brundtland Report that had come to represent the notion of sustainability in terms of 'think global, act local'. Steve Hinchliffe (1996: 56) argues that:

> the 'global' end of this dichotomy was being progressively redefined as the 'scientific composition of the atmosphere'. At the same time, 'the local' was being progressively expressed in ways that resonated with the prevailing political culture in Britain ... [which resulted in] the 'local' being equated with the

'individual' or private consumer. Thus, rather than helping the earth beginning with the 'neighbourhood' or the 'community', it began in the 'home.'

Similarly, the establishment in 1992 of the Energy Savings Trust was intended to provide an independent source of advice and information on energy efficiency in the context of the privatisation of the energy supply industry and the imperative of reducing greenhouse gas emissions, with the individual householder or consumer again at the centre of government attention. Successive campaigns over the past decades, from the Are You Doing Your Bit programme, established in 1998 by the Department for Environment Transport and the Regions, to *Act on CO2*, launched in 2009 by the Department of Energy and Climate Change, have similarly sought to leverage the response of individual households as a means of addressing climate change, accompanied by a suite of tools, mechanisms and forms of measurement that are readily interpreted as a form of governing associated with a neoliberal impulse to harness the governing of the self as a means through which to achieve broader sustainability ambitions (Summerville et al. 2008).

However, beyond these initiatives, parallel developments through the 1990s saw the development of more consciously local and collective attempts to respond to climate change. Through the first part of the decade, local authorities in the United Kingdom embraced the notion of sustainable development and in particular the call for action at the Rio Earth Summit that they should develop their own Local Agenda 21 (LA21) plans to respond to its imperatives. Commentators suggested that this gave rise to creative and innovative responses across the United Kingdom (Agyeman & Evans 1994), though assessments suggested that by 1994 there were approximately fifty local authorities pursuing this approach, only 36 per cent of local authorities surveyed in 1998 had produced an LA21 strategy and the strength of commitment and level of action involved were regarded as relatively weak (Gibbs et al. 1998; Lucas et al. 2003). Nonetheless, committed to an agenda of devolving powers to regions and to local authorities, the new Labour government endorsed Local Agenda 21, with Prime Minister Blair explicitly calling for all local authorities to have developed a strategy by 2000. Whilst over 90 per cent of local authorities had published such a plan by the end of 2000, the nature and content of these strategies were largely unexamined (Lucas et al. 2003). Yet the apparent appetite of local authorities in the United Kingdom for addressing issues of sustainability was a significant driver behind the introduction in 2000 of a new legal power for local authorities to 'undertake any activities that promote or improve the economic, social or environmental well-being of their area or local people', powers that whilst initially designed to meet sustainable development goals also 'greatly extend the scope for local authorities to take action in support of the [United Kingdom's 2000] climate change programme' (DETR 2000: 41). United Kingdom local authorities had begun in the early 1990s to engage explicitly in the activities of transnational municipal networks, such as the Cities for Climate Change programme run by the International Council for Local Environmental Initiatives (ICLEI) and the Energy-Cities network supported by EU funding (Bulkeley & Betsill 2003; Bulkeley & Kern 2006). Yet these networks were largely animated by the work of a few pioneering local authorities, including Newcastle, Leicester, Kirkless and Merton, with a history of engagement around issues of energy poverty and housing retrofit and an interest in local energy generation. In 2000, with the government's growing interest in climate change coupled with the remit for devolution housed in the Department for Environment, Transport and the Regions, the United Kingdom embarked on a specific pilot of the ICLEI programme entitled Councils for Climate Protection, which involved twenty-three local authorities (DETR 2000: 42), whilst municipalities organised a statement of intent to pursue action on climate change, the *Nottingham Declaration*, which voluntarily committed signatories to work towards monitoring and reducing GHG emissions. By 2010, more than 300 councils had signed the *Declaration,* though its popularity was waning. In 2012, it was replaced by Climate Local, an initiative supported by government departments and the Environment Agency to encourage local authorities to match commitments to reducing GHG emissions with measures to improve local resilience, and by 2014 some eighty-four authorities had signed up to this new scheme (LGA 2014).

Whilst those local authorities that joined the early wave of transnational municipal networks active in the United Kingdom in the 1990s tended to represent medium- and small-sized urban communities, during the 2000s the United Kingdom's large and more economically powerful urban authorities began to express their interests in the climate change agenda. In turn, as is evident in other cities across the world, climate change shifted from being considered as an issue largely associated with the voluntary actions of municipalities to one that was of strategic importance to the development of the urban economy (Bulkeley & Betsill 2013; Hodson & Marvin 2009). In London, for example, the creation of the Greater London Authority and the office of mayor saw the election of Ken Livingstone in 2000. Shortly after coming into post, Livingstone established the London Climate Change Partnership, a stakeholder group charged with developing knowledge and guidance to assist London in preparing for the impacts of climate change. In 2004, the Mayor's Energy Strategy established an initial target of reducing carbon dioxide by 20 per cent below 1990 levels by 2010 as a first step to a reduction of 60 per cent by 2050, and in 2007 the Climate Change Plan declared the target of stabilising carbon emissions in 2025 at 60 per cent below 1990 levels. In 2007, a group known as the Core Cities, including Birmingham, Bristol, Leeds, Liverpool, Manchester, Newcastle, Nottingham and Sheffield, signed what was termed a 'shared commitment to climate change action from the Core Cities and Government' signalling their intent to mobilise resources around this agenda (Core Cities Group 2007). Acknowledging that the combined carbon footprint of these cities was some 30 per cent of the United Kingdom's total, the group committed itself to a range of economic and financial measures aimed at supporting the low carbon economy and green growth. Indeed, it has been this agenda of economic imperative and the potential for green growth that has animated much of what has been undertaken in relation to climate change at the urban and regional level in the United Kingdom (Hodson & Marvin 2013).

The 2000s, in addition to witnessing a burgeoning interest in the United Kingdom's cities, also saw a growing engagement with climate change amongst rural communities in the United Kingdom. If during the 1990s the United Kingdom's energy policy and its investment in renewable energy were largely figured in terms of the role of private sector interests, by the early 2000s a new focus emerged in 'both policy discourse and the investment of public resources around the concept of "community" renewable energy development' that moved beyond the usual tropes of consulting with affected publics to include 'notions of community led, controlled and owned renewable energy installation development' (Walker et al. 2007: 65). This growing interest was manifest in a series of reports and documents, including the RCEP's *Energy in a Changing Climate* and the 2003 Energy White Paper. A series of programmes were also set up in order to support the development of community-renewable energy projects. In 2001, Community Action for Energy was established as an advice and networking organisation to promote renewable energy projects at the community scale. In 2002, this was followed 'by the Community Renewables Initiative . . . established in 2002 by the Countryside Agency, which also acts as manager and coordinator of the initiative, with funding provided by the Department of Trade and Industry' to support and develop a range of renewable energy projects across rural England (Walker et al. 2007: 69). A suite of further programmes, including the Scottish Community and Households Renewables Initiative, the Community Energy scheme and the Energy Saving Trust's Photovoltaic programme, rapidly followed suit, with the Clean Skies initiative established in 2003 providing capital funding for projects. By 2004, research found that more than five hundred schemes or projects had been established in the United Kingdom that were in some way framed as involving community energy (Walker et al. 2007). There was, however, little coordination or clear sense of either what engaging the community in this manner might involve or what it might mean (Walker & Devine Wright 2008). Research suggests that the drivers behind this explosion of interest in the community as a site and vehicle through which to realise transitions in the energy system were connected to three main rationales. First, communities felt the need to develop a means through which to counter growing public concerns about the impacts of on-shore wind energy generation. Second, communities held an interest in the potential economic benefits that such projects could bring to rural areas, particularly in the wake of several economic crises to affect the United Kingdom countryside during the late 1990s, including bovine spongiform encephalopathy (bovine spongiform encephalopathy [BSE], or mad cow disease). And finally, the characteristic of community energy as a 'not for profit' enterprise provided 'a route for direct government subsidy rather than the stimulation of the market through more roundabout means' (Walker et al. 2007: 72).

Whilst many of the schemes were relatively short-lived, interest in communities as sites through which energy system innovation can take place has persisted. By 2012, an audit of community-based responses to climate change found a wide array of actors and schemes that were involved in supporting such action:

> Community Energy Scotland has 302 projects . . . there were 24 Carbon Rationing Action Groups active in 2008 . . . Energyshare lists 1087 "groups" (. . . this includes anyone who has registered as being interested in their work) . . . the Buildings Research Establishment's Low Carbon Building Programme included

> 2480 non-domestic, not-for-profit, community based organisations (schools, social housing associations, churches, but not necessarily community energy groups) between 2006 and 2011 ... and Keep Scotland Beautiful's Climate Challenge Fund made awards to 261 communities between 2008 and 2011 (Seyfang et al. 2013: 979).

The United Kingdom government's interest in the community as a means through which to affect responses to climate change has also continued, with specific schemes such as the Rural Community Renewable Energy Fund (DEFRA 2012), the Local Energy Assessment Fund from the Energy Savings Trust and the Low Carbon Communities Challenge, a £10M programme run by the Department for Energy and Climate Change between 2009 and 2011 to provide 'financial and advisory support to 22 test bed communities' across England and Wales (DECC 2012). Such schemes have provided vehicles not only for stimulating the market for renewable energy generation but also as a means through which to encourage individual action on climate change (Fuller & Bulkeley 2013). For example, the United Kingdom Low Carbon Transition Plan, in championing a community-based response to climate change, suggests that 'we often achieve more acting together than as individuals. The role of the Government should be to create an environment where the innovation and ideas of communities can flourish, and people feel supported in making informed choices, so that living greener lives becomes easy and the norm' (HM Government 2009: 92). The focus on community energy was given further impetus with the introduction in 2010 of a feed-in tariff at various rates for microgeneration technologies. Whilst not explicitly targeting community-led renewable projects, many have sought to develop installations on the basis of the financial returns made possible by larger-scale investments and have established new forms of collective organisations, notably mutual or cooperative societies where installations are collectively owned and any income is shared or reinvested in the community.

Whilst the intervention of central government, and the devolved administration in Scotland and Wales, has been critical to the growth of community-level responses to climate change, a range of other organisations has also been critical to this dynamic. Various not-for-profit organisations have sought to mobilise community responses through specific programmes, and the private sector has also engaged in promoting community-level action. In 2008, Nesta launched the Big Green Challenge designed to stimulate and support community-led responses to climate change through selecting ten communities to put their ideas into practice over a year, with four schemes receiving a share of a £1M prize. Similarly seeking to reward success and champion good practice, the Ashden Awards programme provides prizes to exemplary projects and has presented more than seventy awards to community-energy schemes in the United Kingdom to date. A competition-based scheme, Green Streets, was also launched by British Gas in 2007–2008 and a second round of the scheme undertaken in 2010–2011, to 'help communities around the United Kingdom to save and generate energy' (British Gas 2012). Funding was provided to neighbourhood community groups to install energy efficiency measures and renewable energy technologies, with a prize given to the communities that managed to save the most carbon over the life of the project. The Energy Savings Trust's Green Communities Scheme works through a different register, by providing a platform and network through which communities can access advice and support, and in 2011 had more than three thousand members (Fuller & Bulkeley 2013).

Despite the plethora of interventions designed to shape community responses to energy and climate change over the past fifteen years, it has been the grassroots mobilisation of communities themselves that has made the most impact on this landscape. In 2006, the first Transition Town initiative was formed in Totnes, Devon. As the organisation's town's website explains, a Transition Initiative:

> is a community ... working together to look Peak Oil and Climate Change squarely in the eye and address this BIG question: for all those aspects of life that this community needs in order to sustain itself and thrive, how do we significantly increase resilience (to mitigate the effects of Peak Oil) and drastically reduce carbon emissions (to mitigate the effects of Climate Change)? (Cardiff Transition 2014).

From these initial beginnings in the southwest of England, the Transition movement quickly gained momentum across the United Kingdom so that 'by 2008, some 100 communities had joined the network, primarily based in the United Kingdom, but also including the United States, Australia, Japan, and Chile, and by 2014 this number had reached more than 470, with groups also starting in Asia, India, and South Africa' (Bulkeley & Newell 2015: 103). The growth of the movement has been accompanied by a more formal structure, such that a list of sixteen criteria to which prospective groups should comply has been compiled (Mason & Whitehead 2012;

Neal 2013). At the heart of the Transition Towns approach is a focus on the relocalisation of the economy, expressed in relation to various different issues from energy to food, local currencies to transport (Feola & Nunes 2014). Whilst Transition Towns has a specific approach and philosophy, it is symptomatic of a broader engagement by a range of grassroots organisations in the United Kingdom, from neighbourhood associations to church groups and rural and urban communities, with the issues of energy and climate change. In 2009, the Low Carbon Communities network was established to connect these groups and organisations and by 2012 had more than six hundred members ranging in size, scope and scale. Such has been the growth in the organisation and momentum of community-based responses that in 2014 the United Kingdom government published its first Community Energy Strategy (DECC 2014b). Introducing the strategy, Ed Davey, Secretary of State for Energy and Climate Change, argued:

> For too long, community energy has been a policy footnote, with all the focus on big generators and individual households – all but ignoring the potential of communities to play a key role. . . . Already there have been more than 5,000 groups in the UK working to transform how their community uses energy. . . . this Strategy aims to help these existing groups grow and to inspire more to set up and expand (DECC 2014b: 3).

Promising a 'ruthlessly pragmatic' approach, the strategy includes new initiatives for supporting community groups to access information and resources and develop new partnerships with the private sector, and £10M of new funding in the form of an Urban Community Energy Fund (DECC 2014b). Here, as elsewhere across the landscape of climate governance in the United Kingdom and beyond, community is used as a means through which forms of self-reliance and the practices and desires with which it is associated are mobilised in order to accomplish the governing of climate change.

Cases in Accomplishing Climate Governance

The picture that comes into focus through examining the multiple strands of government, corporate and community responses to climate change is one of a complex patchwork landscape, where the conjunctures of particular events, issues, entrepreneurs and intentions have served to create various (sometimes unintended or unpredictable) sites and arenas through which the governing of climate change is conducted. The governing of climate change has not progressed in a linear fashion, as one might expect from the growing scientific evidence of the impacts of climate change or the rising levels of greenhouse gas emissions. Nor has it been a matter of the policies of central government simply trickling down to affect the behaviour of corporations, communities and individuals. Rather, the governing of climate change twists and turns as it is woven through different realms. The task of locating climate governance may seem rather bewildering or overwhelming. If climate governance is to be found everywhere, does it also escape any meaningful definition, account or analysis? As Allen (2004) argues in his essay on the 'whereabouts of power', the complex spatialities of power does not mean that it is without geographies. Rather, what is at stake is the analysis of how power is itself shaped and manifest through its geographies. Likewise, tracing the whereabouts of climate governance becomes not a matter of identifying the particular institutions or territories over which it takes effect, but rather of examining how the orchestration of climate governance is accomplished through complex and interwoven political geographies that draw together what might appear to be distinct elements through, for example, transnational networks or other forms of association that make what appear to be distant objects proximate, whilst at the same time attending to the ways in which seemingly mundane, local or transient political moments can gather momentum and reach (Allen 2011; Bulkeley 2012).

The cases of climate governance with which this book is concerned are all of this nature. They jumble together the near and the far, the mundane and the extraordinary, the public and private, in ways that trouble current understandings of the whereabouts of climate politics. In this manner, whilst based in the United Kingdom, each case has resonance with the forms of climate governance that are emerging in multiple different contexts, and indeed several owe their character to elements garnered beyond the jurisdictional boundaries of the United Kingdom. In the rest of this chapter, each case is briefly introduced in order to provide a context for the more detailed elaboration that follows in the subsequent chapters. Each of the cases was researched over the period 2008–2013, using a range of methods, including interviews with those who designed or participated in each scheme (approximately fifty interviews were conducted across the cases), the analysis of documentary materials (including policy documents, press releases, website content, media coverage, technical guidance and third-party assessments) and participant observation (at meetings, public events and project activities).

Three of the cases take their starting point in the realm of corporate responses to climate change, examining how climate change has come to be regarded as an issue that requires intervention and the means by which governing is accomplished. The cases include two from sectors in which there has been a long-standing engagement with climate change – energy and banking – and one from the retail sector, where climate change has been a more recent concern. Now one of the world's largest banking groups and headquartered in London since 1991, the Hong Kong and Shanghai Banking Corporation (HSBC) was originally founded in East Asia in the middle of the nineteenth century. HSBC was one of the first banking groups to engage in the international debate on sustainability and climate change, joining the UNEP banking initiative in the early 1990s, and in 2004 it signalled its intent to become the world's first carbon-neutral bank. Within HSBC, the response to climate change was catalysed through the formation in 2007 of the HSBC Climate Change Partnership. This programme of activities involved a range of nongovernmental organisations as partners in a suite of activities to engage HSBC staff in developing their understanding of the climate change issue and its implications for business as well as work intended to support scientific assessment of climate change and the preservation of vulnerable ecosystems. This book examines the development of the Climate Change Partnership, focusing on the Climate Champion scheme developed with HSBC staff as a means of enhancing their awareness and making the business case for climate action. In terms of the energy sector, the case explored in this volume centres on the Energyshare scheme developed between 2011 and 2012 by British Gas, the United Kingdom's largest energy supplier and a subsidiary of the energy company Centrica, in partnership with River Cottage, an enterprise established by TV chef Hugh Fearnley Whittingstall that includes both a commercial food business and support for various sustainability issues. Following its experience with Green Streets, British Gas was seeking new means through which to examine the nature and implications of community-based energy schemes whilst River Cottage was seeking a means through which to support a more grassroots approach to energy provision. Energyshare included a novel tariff that offered consumers 100 per cent renewable energy and a donation of £10 to a community energy scheme of their choice, a competition for community energy groups to win funding support for particular schemes and an online platform for sharing ideas and experiences across the community energy sector. The analysis here focuses on how this scheme came to shape the development of climate change responses in one community, Hexham, Northumberland, as detailed later in this section.

A third case of the corporate response to climate change focuses on the supermarket sector. As Gouldson and Sullivan (2012) argue, supermarkets are a critical part of the retail sector, significant because of their economic dominance, the long history of corporate social responsibility initiatives and the recent shift in approach towards climate change that has seen seven out of the nine largest supermarkets in the United Kingdom adopt more holistic approaches to the issue across their businesses. Tesco, the case analysed in this volume, was at the time that this research was conducted (2009–2012) 'the third largest retailer in the world' and one of the first to adopt climate change targets for its own business (Gouldson & Sullivan 2012: 3). This analysis focuses on the ways in which Tesco sought to accomplish the governing of climate change through the design and development of low carbon stores on the one hand and through the implementation of measures intended to enable consumers to reduce their own carbon footprints.

Three of the cases in the book take an alternative starting point, examining the ways in which climate change has come to be made into a community concern. Located to the west of Newcastle in Northeast England, the Hexham River Hydro project has its roots in the autumn of 2010, when Hexham Community Partnership, a local development trust engaged in economic development and regeneration activities, joined forces with the Transition Tynedale Energy Group and the Hexham Town Council to establish a working group to explore the potential for a community-owned hydroelectricity scheme on the River Tyne in the small town of Hexham. Central to the development of the project was the decision to enter the Energyshare competition, described earlier in this section, and the project's success in securing the £100K prize for the large project competition. Whilst the scheme was abandoned in 2013, this analysis examines the processes through which it came be formed, mobilised and politicised as a means through which climate change could be governed. Rather than being fostered through the actions of corporate interests and nongovernmental organisations, the other cases of community-based climate responses were given impetus by the interventions of various government bodies. Also based in the Northeast of England, Berwick is a town of approximately thirteen and a half thousand people near the Scottish border. With its antecedents in a community renewables conference held in the town in 2005 at the initiative of the local strategic partnerships board, Community Renewables (CoRE) was formed as an organisation that sought to facilitate renewable energy projects across the region. Berwick CoRE was formed in 2007 as a partnership between CoRE and the Berwick Community Trust, a local charity established in 2004 with the aim of fostering the local development and well-being of the community. Initially focused on the development and installation of a local wind turbine, Berwick CoRE was selected to participate in the United Kingdom government's Low Carbon Communities programme (2010–2012), which provided a grant of £250,000 to purchase and install 40 PV panels and support community engagement with energy efficiency. Through this case, the analysis examines how these different projects and dynamics shaped the ways in which climate governance was accomplished in Berwick and the case's wider implications.

The final case explored in this book is that of the Low Carbon Zones in London, focusing on the area of Brixton, an inner-city district of London located within the Borough of Lambeth, which has a population of approximately three hundred thousand. The Low Carbon Zones scheme was initiated by the Conservative mayor of London, Boris Johnson, in 2009, signalling his intended commitment to the climate change agenda established in the early 2000s by his predecessor, Ken Livingstone. Under the scheme, Borough Councils were invited to submit applications to designate zones within their jurisdictions that would become the focus for energy saving measures designed to achieve a reduction in carbon dioxide emissions of 20.12 per cent by 2012, a symbolic target intended to make a contribution to the greening of the 2012 London Olympics, with the broader aim that 'the Zones act as exemplar models for a whole-community approach to reducing CO_2 emissions that can be rolled out within and beyond London' (Greater London Authority 2011a: 111). Whilst the scheme was initiated from the top down, central to the practice of governing carbon in the Brixton Low Carbon Zone was the participation of Transition Brixton, a member of the Transition Towns network, alongside a range of other local actors and organisations, from the local borough, community groups and housing associations. The book examines the way in which the Low Carbon Zone was defined, enacted and extended through the practice of governing climate change.

Together, the cases enable an exploration of how and with what implications the governing of climate change is being accomplished. As set out in the previous chapter, attending to the ways in which accomplishing governance takes place involves examining how such forms of governing are realised and attained on the one hand, and how such practices come to be made skilled and proficient. Rather than being focused on what such schemes and initiatives have achieved in absolute terms – such as in terms of the reductions in greenhouse gas emissions or kilowatts of renewable energy produced – this book is concerned with examining the ways in which governing is pursued through their configuration and mobilisation, with the geographies and politics of accomplishing climate governance, and with the wider implications for how we conceive of and intervene in the governing of climate change. Through the next four chapters, the cases introduced are used to explore and illuminate these central questions in detail through consideration of how governing is authorised, ordered, articulated and made public, whilst the concluding chapter reflects on the implications for how we might come to regard the nature and possibilities of climate governance anew.

3

Practising Authority

Introduction

If governing is not a universal condition but rather a specific means through which power is articulated (see Chapter 1), at its heart lies the working of authority – commonly understood as 'the rightful governance of human action by means other than coercion or persuasion' (Litfin 2000: 124; see also Agnew 2005; Lake 2010). In contrast to coercion, where power operates through force, or other registers of power, where appeals to reason and self-interest are made, authority is marked by its reliance on recognition. Actors come to regard institutions, rules and norms as those with which they ought to comply because they are right, proper or appropriate (Bernstein 2011; Hurd 1999; Mol 2008). This is not, however, 'to suggest that relations of authority are freely entered into . . . [they] are structured by uneven political economies which shape the dynamics of how and with what effect authorisation is given' (Bulkeley 2012: 2430). Nonetheless, it is the particular character of authority as the recognised orchestration of power, a mode particularly suited to 'activities that require any measure of creativity or enthusiasm' (Hurd 1999: 385), which suggests that it is central to accomplishing governance.

Indeed, so central is the idea of authority that much of the debate on governance has sought to attend specifically to its shifting geographies, to the ways in which authority appears to be moving above, below and beyond the state, emerging in new actors and partnerships and taking form in private spheres previously considered to be beyond the realm of public goals. Across the field of global environmental governance, it has been these dynamics and the agents through which they are manifest that has attracted the vast body of scholarly attention (Lemos & Agrawal 2006; Newell et al. 2012). Yet despite this enthusiastic engagement with the ways in which authority is and could be made differently in order to govern the climate, its conceptual foundations remain relatively unexplored. As a result, there has been limited engagement with how and why we might want to think about and examine authority differently in relation to climate change (Bulkeley 2012; Jagers & Stripple 2003). Whilst scholars have sought to understand the different basis upon which recognition might be won, from scientific to moral to economic authority, the critical means through which power comes to be articulated in relation to authority remains relatively taken for granted. Underpinning this stasis is what John Allen (2004: 22) terms a 'vocabulary of power' based on the idea of capabilities held within particular bounded units and the 'dispersion or distribution of powers between various levels and sites of authority'. Where power as seen as a capacity of individuals, institutions or particular sets of resources, it comes to be regarded as a finite quantity that can be divided, moved, won or lost as it is held over others in order to ensure their domination, such that authority is determined through the actions of an 'explicit, wilful agency' operating through a 'bilateral, immediate' relation (Conca 2005: 186).

Thinking authority differently requires, then, more than an appreciation of the different agencies and forms of recognition involved: it entails reconsidering the ways in which power is orchestrated and practised (see Chapter 1). Rather than seeing power as a held and finite capacity, other theoretical traditions point towards an understanding of power as 'a general facility for enabling things to happen, where power itself is not conceived as a resource but as something generated by the application of resources and skills over tracts of space and time' (Allen 2010: 2900; see also Barnett & Duvall 2005; Lipschutz 2005) or as an immanent force, constitutive of and constituted through sociomaterial relations such that it is 'inseparable from its effects' (Allen 2003: 65). Despite differences in the ways in which power is conceived within these traditions, together they open up the notion of authority in several important ways. First, if power is mediated and generated through social relations, there is no zero-sum game in which as some actors gain power, others must lose; instead, new arenas of governance are successful precisely because of 'their abilities to generate authority' (Hajer 2009: 22). Second, power itself is not of one kind, but rather takes different specific forms, from domination to seduction, coercion to inducement, each of which has distinctive characteristics (Allen 2003). Moreover, power is orchestrated in different ways. One distinction is that between the instrumental qualities of power, where it is intended to 'bend will', and, after Arendt, the ways in which power emerges through association, creating a 'common will' (Allen 2003: 123). Foucault's work (2009) also signals that power is manifest in different forms, as sovereignty, discipline and government, with each modality invested with distinct purpose. Taken together, these insights suggest that 'authority, as one form of power, could be enacted instrumentally, affected in association, or orchestrated through government' (Bulkeley 2012: 2432), such that the imperative in terms of understanding how governance is accomplished is not one of establishing authority's essence but rather of examining the 'various practices of authorisation' (Stripple 2007: 14) employed through multiple modalities in pursuit of the will to govern.

A third way in which the notion of authority can be reconsidered through such perspectives concerns the ways in which it is practised and comes to be recognised. Commonly authorisation is conceived in terms of the attribution of consent. Yet, as Hajer (2009) suggests, there is more to authorisation than mute acceptance. He argues for an approach that recognises that authorisation is secured through 'reasoned elaboration' such that 'authority is not about submission, not about acquiescence ... it is achieved through communication; it is about the development of a way of seeing things that can be, and indeed is, taken up by others and which results in the acceptance of a particular line of thinking and acting' (Hajer 2009: 22). Thinking recognition relationally in this way allows us to consider how it may be mutually arrived at, through the building of consensus, as much as it may be leveraged through forms of instrumental power to create consent to such shared ways of seeing. Equally, where power is manifest in its governmental guise, recognition may work differently for 'when power operates at a distance, people are not necessarily aware of how their conduct is being conducted or why, so the question of consent does not arise' (Li 2007a: 5). This does not, however, suggest an absence of authorisation, but rather that it works implicitly by placing 'action, practices, symbols, and institutions into a meaningful social frame of context' by subjecting 'something or somebody to the operation of a system of meaning, significance and power' (Latham 1997: 13, cited in Stripple 2007: 15).

Finally, viewing power as constitutive of and generated through social relations raises questions of its spatiality that exceed prevailing accounts of the shifting geographies of authority (Allen 2003, 2004; see also Chapter 1). Rather than being fixed by spatial coordinates or the arc of power carved by particular agencies, the geographies of authority are constituted and (re)produced through the mediated sociomaterial relations within which power is effectuated (Bulkeley 2005; Painter 2010), so that 'power relationships are not so much positioned in space or extended across it, as compose the spaces of which they are a part' (Allen 2011: 284). In this manner, as discussed in Chapter 1, '*spatiality is constitutive of power relations not only in general, but also in the particular ways in which different modes of power take effect*' (Allen 2003: 102, emphasis in the original). For Allen and Cochrane (2010), new forms of topological power are critical to the contemporary working of the state, such that conventional notions of time and distance are rearticulated through practices of proximity and reach that draw together different constituents into various assemblages through which power comes to be ordered and experienced. Rather than viewing the analytical task as being one of attending to the shifting patterns of authority, the question becomes one of the ways in which 'different socio-spatial relations are constitutive of different modalities of power and how these processes in turn serve to sustain distinct spatial orderings (such as the local, community, territory, the transnational and so on)' (Bulkeley 2012: 2433). Exercised through recognition, authorisation requires that forms of connection and proximity are established, even where geographical distance may be significant, in turn serving to constitute the forms of assemblage through which governing is accomplished.

Rather than considering authority as one form of power with multiple sources of recognition (e.g., moral authority, scientific authority, personal authority), this analysis suggests that authorisation can be practised through distinct modalities 'constituted through the purposes for which power is deployed, the nature of recognition and compliance, and distinct forms of socio-spatial relations' (Bulkeley 2012: 2434). From the starting point of the instrumental, associational and governmental ways in which power might be conceived, three distinct modalities of authority can be identified – as consent, consensus and concord (Bulkeley 2012). Given that 'authority is not an achieved condition ... but a complex and variable practice' (Stripple 2007: 15), these are necessarily ideal types, intended to open up the ways in which authorisation can be conceived and the analysis of the messy actualities through which it comes to be articulated and contested.

It is to the practice of authority that the rest of this chapter turns, examining the ways in which governing climate change is authorised in the range of different sites and spaces of climate governance that have emerged in the United Kingdom over the past two decades. Whilst the cases with which this book is concerned vary in many ways, from the boardrooms of international corporations to the banks of local rivers, analysis suggests that they have much in common in terms of how authorisation has come to be practised. Although, as the second section of the chapter explores, there is evidence that securing consent has been important to the evolving governance of climate change, particularly in relation to the work of central government, it has been through consensus that much of the governing of climate change has been accomplished. For some, the dominance of consensus in climate politics is of significant concern, signalling the post-political condition of our time and suggesting an absence of dissent from or resistance to dominant elites (Swyngedouw 2010). Yet the analysis of the practice of consensus undertaken in the third section of this chapter suggests that politics is far from absent, but rather takes different forms as resistance and contestation continually emerge alongside the working of consensus. Authorisation as concord, through 'the "bringing into agreement" of heterogeneous elements – social and material – in such a manner that they are taken for granted as authoritative in the shaping of conduct' (Bulkeley 2012: 2439), is less clearly visible through the reading of the cases presented in this chapter. Authorisation requires 'the complex assemblage of diverse forces ... such that aspects of the decisions of individuals, groups, organisations, and populations come to be understood and regulated in relation to authoritative criteria' (Rose & Miller 1992: 183). Accomplishing governance in this manner requires ongoing governmental work and the deployment of a range of techniques and practices explored further in Chapters 4 and 5.

Consenting to Climate Governance

As argued previously, central to the way in which authority has traditionally been conceived is the idea that it requires the consent of the governed in order to be legitimate and maintained in particular institutional configurations, such as the state or a political party. Those concerned with the ways in which non-state actors are able to garner legitimacy and govern different kinds of constituencies suggest that this comes through demonstrating their expertise, moral position or market prospects (Cashore 2004; Rosenau 2002). In its consent mode, practicing authority requires the ability to enlist 'voluntary agreement' or 'acquiescence in what another proposes or desires' (*Oxford English Dictionary* 2015). As consent, authorisation is 'designed to make others act in ways that they otherwise would not have done' (Allen 2003: 117). Critically, rather than such forms of authorisation being held by particular actors or institutions, as in its its other guises, authority as consent is generative, taking effect only when it is enacted and found to be 'worthy of acceptance' (Friedrich 1958: 35, cited in Hajer 2009: 22). In other words, achieving consent in practice requires then not only that a set of mechanisms and techniques is put into place to persuade constituents to acquiesce to particular forms of governmental intent, but also that this is regarded as legitimate and worthy of compliance. Authority as consent is therefore relational: formed in the working through of techniques for generating acquiescence and the practice of compliance. For those engaged in complying, such forms of consent can either serve to keep powerful interests at bay, avoiding the imposition of more coercive modes of power, or provide a means through which they can achieve their own ends. In the case of voluntary certification of carbon emissions, for example, compliance has enabled firms to secure various forms of corporate social responsibility whilst also contributing to the development of new economic domains, including carbon trading and different forms of carbon investment. Consenting to authorisation not only conveys legitimacy to those who seek to govern, but also to those who are subject to being governed.

In the arena of transnational climate governance, consent-based authority takes place in those arrangements that rely on mechanisms of certification or standard setting and that sought to enrol constituents in undertaking these forms of regulation and compliance (Bulkeley 2012). Yet despite the prevalence of such forms of authorisation in the international and transnational spheres, one of the intriguing aspects of the landscape of climate governance in the United Kingdom (see Chapter 2) is the relatively limited use of forms of consent outside of the practices of the national government. Whilst the development and implementation of a series of regulative measures – from the 2001 Climate Levy to the 2008 Climate Change Act – have been central to national climate policy over the past two decades, beyond the national government arena forms of consent-based authority appear a relatively rare occurrence. Of course, many

UK-based entities and organisations have participated in the design of transnational governance arrangements through which such forms of authority are generated – such as the Carbon Disclosure Project and the Voluntary Carbon Standard and Gold Standard for carbon offsetting. Yet the techniques that have been developed and assembled in order to govern climate change in the United Kingdom beyond the bounds of the national regulative apparatus appear to make more limited use of the practice of authorisation through consent. One partial exception has been the development of community-based climate change initiatives. As Chapter 2 outlined, over the past fifteen years the idea of a community response to the twin challenges of energy and climate change has come to take a central place in grassroots organisations, national government and some private sector organisations. In the remainder of this section, the ways in which such programmes and initiatives have been realised through the practice of consent are explored. Compliance, and hence the generation of authority, has been achieved through the use of consultants and other external agents that enable community-led initiatives to acquiesce to the intentions of governmental programmes whilst continuing to put climate change towards their own ends. At the same time, the cases suggest that what constitutes compliance is far from concrete, as the ends towards which communities are being enrolled are reconfigured and reworked over time.

Community Compliance?

The introduction of a suite of programmes to support community-led energy and climate change initiatives in the 2000s has led to the development of several hundred groups and projects across the United Kingdom (Seyfang et al. 2013). Following a series of programmes specifically aimed at development microscale renewables and some energy savings measures in the early 2000s, a more explicit climate change focus can be discerned in the programmes that were initiated in the latter half of the decade (see Chapter 2). In England, the Department of Energy and Climate Change established the Low Carbon Communities Challenge, a £10 million programme that provided grants and advisory support to community groups for the installation of low carbon technologies, infrastructure and behavioural change initiatives (Catney et al. 2014; Creamer 2014). In Scotland, the devolved administration established the Climate Challenge Fund in 2008 with the intention of empowering ‘Scottish communities by building the capacity for sustained future [carbon] reductions' (Gunn 2008: 3; quoted in Creamer 2014: 4). The Fund was extended until March 2016, and it is estimated that 'almost £70 million will have been distributed to hundreds of carbon-saving, community-led projects, enabling activities such as local recycling collections, construction of micro-generation infrastructure, electric vehicle trials, provision of community allotments, and completion of home energy audits' (Creamer 2014: 4). In the same period, the incoming mayor of London, Boris Johnson, established a Low Carbon Zone programme in the Greater London area through which communities were to collectively reduce their carbon footprints by 20.12 per cent by 2012 (see Chapter 5). Grant funding has been vital to the development of community-based climate change and energy projects in the United Kingdom, with a survey finding that it was 'clearly the major source of financial support [for] these projects ... 72% had been successful in winning grant funding, while a further 10% had applied but been unsuccessful' (Seyfang et al. 2013: 984).

Such funding mechanisms not only have served to establish community as a critical site for climate governance, but are a key means through which the practice of consent is orchestrated. Common across these programmes is the intention that community-based initiatives should deliver real emissions savings – savings that would be measurable and verifiable. In the case of the Climate Challenge Fund, it is 'a requirement ... that groups deliver measurable CO_2e [carbon dioxide equivalent] reductions within the lifetime of the funding, most commonly one year' (Creamer 2014: 13). To this end, the implementing agency, Keep Scotland Beautiful, is clear in terms of the criteria that those seeking funding will have to meet:

> Your first outcome should state the CO_2e reductions your project is aiming to make and what you are going to target to make that reduction ... [and] you should identify the CO_2e baseline against which you will measure your CO_2e reduction (KSB 2013: 11; quoted in Creamer 2014: 8).

In London, too, the ten communities selected as Low Carbon Zones needed to establish a baseline of emissions and measure and verify that reductions in the order of 20.12 per cent had been achieved over the lifetime of the project. The impact of dependency on external funding by community groups is well documented, including administrative burdens, the impact of project funding cycles on the long-term development of projects and the shifting nature of agendas in order to comply with the demands of external agents, all of which have been documented in relation to community-based energy and climate change initiatives (Seyfang et al. 2013). Furthermore, the dominant framing of addressing climate change as a matter of the verifiable reduction of carbon has also structured the means through which community-based initiatives have come to address the issue on the ground. In Scotland, an evaluation of the Climate Challenge Fund found a dominance of technical measures to improve building performance over behavioural change, at least partially due to the need to demonstrate emissions savings over the short term which in turn had led to groups 'over-prioritising quick carbon savings at the expense of other sustainability goals' (Brook Lyndhurst & Ecometrica 2011: 114; cited in Creamer 2014: 5).

Across these different arenas, the generation of consent with programmes that seek to demonstrate that communities provide an arena where targets for carbon emissions can be achieved has led community groups not only to acquiesce to their work being monitored and verified in carbon terms, but to undertake particular forms of initiatives that will enable them to comply with the requirements to which they have consented. Yet generating this form of compliance is no small matter for these kinds of groups. One interesting outcome is that for many such groups, compliance, and hence the generation of consent, is being outsourced. In Brixton, London, the Low Carbon Zone, a coalition of interests across the local authority and grassroots organisations in the borough, have brought in consultants to ensure that they comply with the funding requirements:

> We've got external consultants working with bringing in the carbon monitoring, and finding expertise on that front. So we've got quite a robust baseline for carbon emissions and monitoring that (Brixton Interview A 2011).

In Hexham, where the Hexham River Hydro project won funding through the Energyshare competition (see Chapter 2), this work of outsourcing was particularly onerous, given that the project needed not only to demonstrate its carbon credentials but also to comply with a whole host of other regulative requirements in terms of environmental impact, engineering standards and energy market regulation:

> If we thought the job was done when we won the Energyshare funding we would have been very disappointed! We have continued to work extremely hard over the last few months, albeit largely 'out of sight,' supported by high level advice from both Energyshare and Carbon Leap Frog. However it has meant that Hexham River Hydro can now take a big step forward with the appointment of our consultant team. Following national advertisement, Cundall Johnston & Partners

LLP trading as Cundall, a world renowned consulting engineering practice whose head office is based in Newcastle, has been appointed as lead consultant. Led by Damian Dungworth, they have assembled a multi-talented team drawing on staff members working alongside key specialists. Mann Power Consulting Ltd, based in York will be the key advisor on hydrology, guided by Fishtek a specialist fisheries consultancy together with local ecologist Ruth Hadden from Ryal Soil and Ecology(Hexham Community Network 2013).

Whilst the mobilisation of the project as one that was feasible and could attract momentum within the community required the expertise garnered through the local community organisations (see Chapter 4), authorising the project as one that could comply with the requirements of the funders and regulators relied upon a different set of mechanisms. Consent is not simply given. Nor does it exist before interventions take place. The generation of consent is then an ongoing process. The enrolment of community-based groups into governmental programmes requires an initial acquiescence to the ways in which funding bodies define communities and their role in generating carbon emissions reductions. An ongoing suite of practices is also needed through which enrolment is maintained both in the design of interventions that are able to realise these reductions and in the techniques by which they can be measured and verified.

At the same time, ensuring the practice of consent is a matter not only for the community organisations involved, but also for the funding organisations. The case of Berwick, where community-based organisations sought to develop both renewable energy and behavioural change programmes within the town, partially through the support of the Low Carbon Communities Challenge but also with funding from other regional and European bodies (see Chapter 5), suggests that this practice is also far from assured:

. . . there are challenges in every area of our work, but the climate change work is not well understood by funders. Our experience would be that they're not clear about what they're looking to achieve and that programmes aren't well defined (Berwick Interview A 2011).

This lack of clarity was felt to exist for several reasons. Alongside the churn of staff within funding agencies, a lack of community expertise in the energy and climate change arena meant that there was limited learning within funding agencies as to what good community-based projects would look like and how compliance with policy objectives could or should be achieved:

> Some of that is about the technical nature of some of these subjects. There is a lack of expertise in a community level in tackling it, which means that doesn't feed back up the pipes so that funders start to develop. And I don't think there's a body of good practice, at least that's well documented. There is good practice out there but it's disjointed and doesn't appear well documented. So, funders aren't clear at all (Berwick Interview A 2011).

At the same time, much of the work that communities undertake was seen as inherently difficult to monitor and verify. This applies not only to what were sometimes regarded by funders as additional benefits but which often formed the core rationale for the groups undertaking such work, such as developing a sense of community, improving the affordability of energy services, saving money or generating new forms of local economy, but also to climate change measures themselves, particularly those that sought to target the actions of individuals:

> I think the other thing about this kind of work, as well, is that particularly the focus here has been about attitudinal change, right? That's really difficult to generate hard evidence and to demonstrate your outputs is challenging. So ... and in their defence, maybe that's why some of the funders have been slow to act (Berwick Interview A 2011).

As a relational practice, authorisation via consent requires not only that legitimacy is established but that the acquiescence with the intention of governmental programmes is secured and maintained. As UK government and corporate bodies have come to regard communities as a site through which climate change ambitions can and should be realised, programmes have been devised through which to enrol such constituents in the work of accomplishing climate governance. Realising such programmes has required the practice of consent, whereby community groups seeking funding must comply with governmental ambitions, particularly those to demonstrate measurable carbon reductions, and those engaged in the orchestration of such programmes are required to actively monitor and ensure compliance. In the practice of consent, however, community groups are able to manoeuvre, finding third parties through which this work can be undertaken, such that the work of consenting to governmental programmes comes to be undertaken through (a relatively small number of) consultancies. Further, as projects evolve, the burden of ensuring compliance becomes challenging for governmental programmes, whose expertise in the area is limited and where multiple benefits are hard to quantify or capture. Authorisation as consent is far from imposed from above, but rather constituted, generated, in the workings of governing climate change in concert with a host of other intentions. What then comes to be legitimate in climate terms is broadly defined and often only loosely accounted for, despite the considerable investment being made in community-based responses.

Creating Consensus

In its associational mode, authorisation relies upon the 'power to connect, to bring together but not to suppress the interests and differences that commonly divide' (Allen 2003: 123). Rather than seeking consent, authorisation here requires the forging of consensus such that those engaged in the will to improve 'seek to secure constituents around common positions and translate their efforts into standardised measures of benefit' (Bulkeley 2012: 2437). In this guise, authorisation is mutually constituted through the association of constituents with one another as heterogeneous interests are reworked into a common position on the nature of the problem that requires intervention and the means through which this should be achieved (Smith & Fichlein 2010). This requires that constituents are bought into proximity and a variety of techniques – from shared declarations of intent, highly visible events, slogans and publications, to the circulation of ideas and experience – are used to this effect, reducing geographical and political distance such that the articulation of any one constituent stands for the rest. Creating such forms of consensus is not without dissension. The process of determining what constitutes the climate problem around which consensus should be forged is fraught with tension, whilst the ongoing work of practicing improvement is continually reworked as this consensus shifts. What is critical, however, is that such contradictions are managed 'less by technique than by compromise in its dual sense: parties to the assemblage make compromises and, in so doing they become implicated, their positions compromised and their critiques contained' (Li 2007b: 279).

Ensuring consensus therefore requires an ongoing process of mediation and negotiation through compromise in its dual sense so that whilst 'constituents may not deliver on specific outcomes ... [they] continually justify their intentions to do so in order to remain party to the consensus at hand' (Bulkeley 2012: 2438). The domain of climate governance has been particularly marked by consensus politics, from the development of international scientific assessments to the emergence during the 2000s of a broad and diverse range of interests supporting action to address the issue (see Chapter 2). Nowhere was this perhaps more apparent than in the United Kingdom, where businesses from traditional fossil fuel companies to major retailers, political parties of all hues, development charities, traditional conservation bodies, environmental campaigners and community-based organisations collectively championed the need to act on climate change. Yet whilst such consensus politics has attracted critique for its post-political nature (Kenis & Mathijs 2014; Rice 2014; Swyngedouw 2010), the actual workings of consensus, its significance in authorising the will to improve climate change and the forms of contestation it provokes have received rather limited attention. In the rest of this section, the ways in which consensus has been practised and disputed are examined in relation to the development of corporate climate change responses, through the case of the development by HSBC of its Climate Partnership, and the emergence of London as an exemplar of low carbon urbanism.

Better Together? HSBC and the Climate Partnership

If during the 1990s a range of fossil fuel interests had sought to deny or derail the governing of climate change, by the mid-2000s a growing number of corporations had come to see climate change as both a risk and opportunity for their business (see Chapter 2). Intriguingly, rather than seeking to influence the actions of government at either the national or international level, this positioning of climate change as a business challenge meant that several leading international corporations came to regard themselves as part of the means through which the improvement of the condition of climate change should be governed. In 2004, at the 10th Conference of the Parties to the UNFCCC in Buenos Aires, HSBC declared its intention to become 'carbon neutral', only the second major financial institution to do so after Swiss Re. Achieving this in 2005, 'it became the first bank and the first FTSE 100 company to become carbon neutral ... by increasing energy efficiency, reducing business travel, and purchasing renewable energy; remaining emissions were offset by purchasing verified carbon credits' (Climate Group 2014b). Whilst clearly imbued with significant symbolic intent, the declaration of carbon neutrality at HSBC was part of a broader assemblage of artefacts and devices through which the government of climate change was orchestrated through the body corporate. Crystallised in 2007 as the HSBC Climate

Partnership, a five-year programme (2007–2011) amongst HSBC, the Climate Group, Earthwatch, WWF and the Smithsonian Tropical Research Institute with an aim to 'inspire action on climate change' (Climate Group 2014c) internationally, with intervention focused on Brazil, China, India, the United Kingdom and the United States, the primary locations for its business:

> Our aims were ambitious – to impact on individuals, governments, the environment and species across four continents. Our targets were a combination of advances in scientific understanding; measurable impacts on the physical environment; improvements in economic conditions and changes in behaviour for millions of people; and accelerated adoption of new technologies by governments around the globe (HSBC 2010: 5).

As the preceding quote suggests, rather than being confined to an external threat, a matter of efficiency or one of corporate responsibility, the problem of climate change was expansive within HSBC, coming to figure in relation to the ways in which HSBC conducted itself both within and beyond the body corporate. Central to the work of governing climate change within HSBC was the formation of the Climate Partnership, casting the 'problem' as one of the need for action beyond the nation-state. Whilst HSBC had previously worked with Earthwatch and WWF on its corporate social responsibility programme, the Climate Partnership marked a departure from business as usual in two important respects. First, the intention was to shift the work of corporate responsibility beyond the boundaries of the organisation, upstream to the sites and arenas through which climate risks were being created (e.g., through work with global cities on reducing their carbon footprints) and downstream to areas of the greatest climate impact (e.g., forests and major river systems). Second, the Partnership placed much more emphasis on collective action as central to addressing climate change:

> . . . the original programme had been HSBC funding the work of three environmental partners including WWF and Earthwatch, who have continued as partners in the Climate Partnership. What we wanted to do was to explore the possibility of getting much more collaboration. If we could get NGOs working together with us, with HSBC, around a particular theme that seemed like a good idea, worth exploring (HSBC Interview A 2010).

This approach, emphasising collaboration and collective action, increasingly became part of the DNA of HSBC's business approach towards sustainability, partly because the shift from local matters of conservation to global environmental change meant that 'a single business would find it very hard to ... own climate change as an issue' (HSBC Partner Interview A 2010). Seeking to garner collaboration in this manner required HSBC not to seek the consent of its partners, but rather to develop a consensus in terms of the nature of the problem and how it might be addressed. Critical to garnering this consensus was the alliance formed between HSBC and the Climate Group through the Partnership. Formed in 2004, the Climate Group 'brought together a handful of leaders and innovators from governments and businesses around the world', a 'group of pioneers' who 'were the first to see the positive business case for climate action' (Climate Group 2014d). At the heart of their work has been the framing of climate change in terms of the opportunities it affords and the development of a suite of programmes and techniques through which it has proactively sought to forge consensus between governments and corporations. In the United Kingdom, the work of the Climate Group was actively supported by Downing Street and former prime minister Tony Blair (see Chapter 2). In 2006, Blair held a summit of business and civil society organisations to promote action on climate change, with the result that the Climate Group was asked to continue this process of convening a consensus for action:

> ... we were asked to go away and think about how can we work with all these other organisations, I say 'we', I hesitate – how do we work with these other organisations to help [them] work together. And that's essentially where the Together campaign came from, which was actually a range of different corporate and civil society partners thinking about how can the government, business and civil society work together? To help the consumers fight climate change. So we were adamant at the time that it wasn't about lists of things to do; it wasn't about really pushing this 'do it to save the planet' finger wagging, telling you what you can't do and what you can do, but it was all about simple, affordable, effective ways that you can save money and reduce your carbon footprint (HSBC Partner Interview A 2010).

Climate change has come to provide a means through which HSBC has reconfigured its role and relationships with a range of governmental and nongovernmental organisations, to become part of a wider group of actors mobilising to address climate change. Aligning themselves through this coalition, HSBC was able to draw on the authorisation of being part of a consensus making the business case for climate change to assemble the climate problematic within the organisation as one not of risk but of economic opportunity. In conjunction with the formation of the Climate Partnership in 2007, Sir Nicholas Stern was appointed as climate advisor to HSBC. Lead author of the 2006 report prepared for the UK Treasury setting out the economic risks and opportunities of addressing climate change, Stern was known for his advocacy of the economic potential of a low carbon transition:

> HSBC wanted to link sustainability and business and around same time Stern report came out [and] Nick Stern is HSBC's climate advisor. I think they very much took that to heart at the senior level about this is how we want to do business, so, yes, we're going to adopt stronger sustainability principles now (HSBC Partner Interview B 2010).

Within HSBC, this economic rendering of the climate problem came to be aligned through the existing practices of the corporation such that a form of climate business has become discernible as a discrete part of the operation of the corporation, in which HSBC's analytic skills and capacities are used to identify 'business solutions that reduce carbon emissions or enable society to adapt to climate change [and] bring environmental and social benefits as well as commercial return' (HSBC 2013: 12). The practice of authorisation here involves not only HSBC working to build coalitions that are party to a consensus about the economic opportunities of responding to climate change, but also serves as a means through which to authorise climate change as a legitimate part of business as usual within the Bank itself.

Blurring the Boundaries: Generating Climate Consensus in London

> . . . climate change is now the centre of political debate in London. In a way even more so than it is in national discussion . . . when Boris Johnson announced [his] candidature for Mayor, even though he's got a record of opposing Kyoto, he's pro nuclear anti wind farms, he still felt he had to say he was going to be the greenest candidate. Because he couldn't let go of that territory and expect to win the election (Interviewee London, 2008, cited in Bulkeley et al. 2014).

The formation of the Greater London Authority (GLA) and the election of Labour politician Ken Livingston as the first mayor of London in 2000 gave renewed emphasis to the need for London to address climate change. Whilst individual borough councils across the Greater London area had been engaged with the issue over the previous decade, in London, as across the United Kingdom, urban responses to climate change remained rather muted, constrained by the limited powers of local authorities and conflicts encountered with agendas of growth and development (Bulkeley & Betsill 2003; Bulkeley & Kern 2006). During the first mayoral term (2000–4), Livingston and his office focused both on the development of the London Climate Change Partnership to assess the impacts of climate change on the city and on devising an energy strategy that placed mitigating climate change at its heart. In line with the main direction of national policy, it argued that 'London should reduce its emissions of carbon dioxide by 20 per cent,relative to the 1990 level, by 2010 as the crucial first step on a long-term path to a 60 per cent reduction from the 2000 level by 2050' (GLA 2004: x; see also Bulkeley & Schroeder 2008). Securing authority for the approach and targets in the plan relied primarily on traditional techniques of creating consent to and compliance with new requirements to provide the on-site generation of renewable sources of energy generation through the planning system.

In contrast, the second term of Livingston's office marked a new approach, whereby the mayor 'decided in the 2004 election that [climate change] was going to be the biggest issue of the second mayoral term, it was the thing that become a personal priority for the mayor' (London Interview A 2007). Authorising climate governance came to revolve around the development of a consensus that climate change was central to the political economy of London, and where London was seen as central to the orchestrating the will to improve climate change. Achieving this consensus required a shift in the terms of debate such that global cities such as London came to be regarded as important strategic actors in the response to climate change (Bulkeley & Betsill 2013). Practicing authorisation in this mode required the building of London's international profile and networks (see also Hodson & Marvin 2007, 2009). Working again with the Climate Group, Livingston and Gavron established the original C20 network of global cities at a summit to profile the role of urban responses to climate change to coincide with the United Kingdom's 2005 presidency of the G8 and the Gleneagles Summit, where national leaders met to discuss the issue. With support from the Clinton Climate Initiative and other foundations, the network quickly grew into the C40 Cities Climate Leadership Initiative and has become a critical means through which the idea of climate change as an urban problem with urban solutions has been made manifest.

At the same time, establishing the consensus that London should and could respond to climate change required the formation of coalitions within the city. At the heart of this approach was the development of the 2007 London Climate Change Action Plan and its ambitious policy goal 'to stabilise CO_2 emissions in 2025 at 60 per cent below 1990 levels, with steady progress towards this over the next 20 years' (GLA 2008: 19). Such an ambitious target may at first glance appear an unlikely means through which to forge consensus, particularly given that by its own admission, the Action Plan recognised the 'difficult truth is that in preparing this action plan we have been unable to present any realistic scenario in which we can achieve the 2025 target set out above, without major national regulatory and policy change' (GLA 2007: 19). Rather than accruing around these targets per se, central to the making of consensus and the practice of authority in London was the critical policy goal of decentralising energy generation:

> The Mayor's top priority for reducing carbon emissions is to move as much of London as possible away from reliance on the national grid and on to local, lower-carbon energy supply (decentralised energy, including combined cooling heat and power networks, energy from waste, and onsite renewable energy – such as solar panels). . . . The Mayor's goal is to enable a quarter of London's energy supply to be moved off the grid and on to local, decentralised systems by 2025, with more than half of London's energy being supplied in this way by 2050 (GLA 2008: 105).

Securing London's energy future came to be a means through which diverse groups, across business and government and between the different political parties and interests at stake, could coalesce. On the one hand, the mayor and the GLA were reliant on the involvement of the business sector to take this strategy forward, for, as one stakeholder reports, 'it's hard to see how in London you can deliver action on sustainable energy. ... having statutory requirements to do the strategy [is one thing] ... in terms of implementation, it's quite clear ... there has to be action elsewhere' (Interviewee, London 2009, cited in Bulkeley et al. 2014: 1479). At the same time, powerful business interests, notably in the form of the lobby group London First, came to support the general direction of the mayor's policy approach whilst also contending its specific direction and implementation. Rather than reject the ambitions of reducing emissions and increasing renewable energy generation, they appealed that they 'welcome the opportunity to work with the Mayor and his team on ensuring that the climate change objectives are met whilst still encouraging development' (London First 2006, cited in Bulkeley & Schroeder 2012: 715).

The mutual dependency between business and industry found across London's politics (Thornley et al. 2005) found expression in the climate change arena in the consensus that a low carbon economy would be good for London's growth, and hence that London should play a strategic role in its development both locally and globally. This logic of a low carbon, decentralised energy regime and its continual authorisation through strategies, programmes and techniques such as funding schemes, community champions, carbon calculations and so forth has served to maintain consensus across the shift in political allegiances that occurred with the election of Mayor Boris Johnson in 2008. As the quote at the start of this section suggests, Johnson was an unlikely figure to champion the climate change cause, yet since coming to office he has 'continued to promote the vision of a low carbon, "green" energy future for London, for example, extending schemes for home energy retrofitting, electric vehicles and decentralized energy' (Bulkeley et al. 2014: 1480). Yet the presence of consensus should not here, or elsewhere, be equated with ready agreement or the evacuation of political contestation over what, and for whom, accomplishing climate governance entails. In London, the apparent unity forged between interests as diverse as London First and Transition Towns, major investors in the city and local community groups, 'belies the diverse rationales at work, conflicts over how and by whom new forms of urban energy should be generated, and the practical and material ways in which low carbon is enacted and disrupted' (Bulkeley et al. 2014: 1484). The practice of authorisation through consensus is an ongoing and unfolding process in which the parties involved seek to exercise compromise but are in turn compromised through being brought into agreement – to fit with one another in ways that might produce unexpected junctures and outcomes (Li 2007b).

Conclusion

Authority, or more accurately for the purposes to which it is put here authorisation, occupies a rather curious place within the study of governance. For some, it is an assumed and yet silent quality of institutions, actors or instruments that would seek to govern. It is tautologically present through the recognition given to particular agents and ways of acting in the world. Elsewhere, much attention has been given to how such forms of authority are created through the qualities themselves that amass authority – expertise, moral position, financial might and so on. Despite their very different levels of interest in the matter of authority, these different views of authority share an underlying sense that it is a property that is divisible, that it can be allocated to particular parties and, perhaps most importantly in the arena of climate change governance, that its whereabouts are shifting between different levels and arenas as the kinds of institutions we call government lose their capacity to be authoritative in relation to international institutions, local bodies, private actors and transnational coalitions. If these are some of the ways in which authority has been treated within the global environmental governance debates, it has fared little better in the emerging scholarship that draws on more critical traditions of social science to analyse climate politics. Though ostensibly concerned with the workings of power and the ways in which these shape the governing of climate change, much of this work has yet to engage with the different modalities of power that might be involved in addressing the particularities of the climate problematic. Authority, in its classical sense of recognition of the governed, appears to have little place in the conceptual repertoires of those bodies of work that seek to collapse the distinctions between actors and networks, the holders of power and those over which it is wielded, society and the material realm and so on.

Yet as Li (2007a: 12) reminds us, even where power is conceived as a relational and immanent quality of sociomaterial conditions, as it is in Foucault's work amongst others, authorisation – what is recognised and what is ignored – is critical. Indeed, she argues, following Rose, that in contrast to the exercise of sovereign power, the particularity of governing is 'to be condemned to seek an authority for one's authority' (Rose 1999: 27). In other words, that authorisation, its practice and contestation, is central to the work of governing. This shift in perspective opens up important new avenues of inquiry. Rather than ask who is governing climate change and how the authority to do so is shifting between the public and the private, the global and the local, we should instead be concerned with how authorisation is practised, how authority is maintained and the ways in which it is troubled and falls into disrepair. Moreover, turning to analyse the modalities of power at work in the practice of authority suggests that authorisation itself can come in different guises – considered here in terms of consent, consensus and concord. Recognition is manifest in multiple ways as power works through different kinds of intervention, sociomaterial relations and technique.

Examining the landscape of UK climate governance over the past two decades, it is possible to trace the outlines of these different kinds of authorisation at work. Critically, the form of authorisation most associated with its conventional definition and interpretation – consent – appears to have limited purchase. Where consent is practised, compliance is far from straightforward with, in the case of community-based responses examined here, forms of outsourcing used in order to ensure that governmental objectives and community ambitions can be pursued in tandem. In contrast, forms of consensus appear widespread. Viewed as a matter of authorisation, as essential to the working of governmental power, consensus can be regarded as distinctively political. Consensus is accompanied by compromise in its 'dual sense' (Li 2007b). Its dominance may be less about the foisting of technocratic intentions by those in power over those who must comply than by a deep ambiguity as to what might constitute improvement in the climate domain and how far climate should come to dominate or compromise other intentions. Here, consensus works to share risk and responsibility, to narrow the terms through which climate can be considered a problem, and to define the field into which legitimate climate interventions can tread (and those that must remain untouched). The result has been one of fitting climate into existing modes and forms of governing, whether through the body corporate or the body politic, which at the same time has served to draw a host of entities into the climate consensus such that they become compromised by the will to improve climate change conditions.

Though consensus appears widespread, settling this matter and ensuring that governing the climate is routine have been far from assured. Consensus as a practice of authorisation is an ongoing process, an accomplishment in the making that requires continual work and compromise. Consensus, both in terms of what the climate change problem is and how it should be addressed, is both precarious and recurrently revisited. New interventions are proposed, the boundaries of the field shifted, alignments recalibrated. Establishing concord, bringing climate change into forms of agreement, is difficult to accomplish across the diffuse sociomaterial relations where climate governance now takes place, requiring both that new forms of order are established in which climate change comes to be comprehended in new ways (see Chapter 4) and calculated and commensurated through the configuration of climate assemblages (see Chapter 5).

4

Bringing Climate Change to Order

Introduction

Climate change has the curious quality of seeming to be everywhere and nowhere simultaneously. As researchers and policy makers have found, the invisibility of the changing composition of the global atmosphere and the intergenerational timescales involved in many of the potential impacts of climate change have meant that it has been hard to grasp as a political and public issue. Yet at the same time the imagery of climate change can appear pervasive. Across the media, urban streetscapes, supermarket aisles and workplaces, injunctions to turn down, switch off or conserve are often accompanied by the symbols of the globe at risk and the rallying cry to do your bit to contribute towards addressing this challenge. Closer inspection reveals of course that climate change is neither absent nor ubiquitous, but rather has come to be related to specific objects, moments and conjunctions. Some are perhaps rather obvious; from energy systems to modes of transportation, houses to whole cities, the moniker 'low carbon' has become attached to a range of entities thought to be critical to achieving reductions in greenhouse gas emissions. Others may appear more specific or more mundane. Take, for example, the Climate Week 2014 campaign, coordinated by a United Kingdom social organisation and supported by several private sector organisations, which encouraged people to 'green your cuisine' by participating in a 'Meat Free Monday' and using low carbon recipes (Climate Week 2014). Another example is the Low Carbon Farming project run by the Soil Association with funding from Defra and the Ashden Trust to encourage their community to adopt new measures to save carbon through the farming lifecycle with a series of programmes, information packs, apps and assessment tools (Soil Association 2014). In contrast, domestic gardening, beyond the extensive discussions of the merits or otherwise of local food production, has to date remained largely outside the scope of climate change, together with many other aspects of our everyday lives and business practices, from going to the gym to eating out, holding workplace meetings and lighting iconic buildings.

Understanding why, how, by whom and with what consequence some parts of the sociomaterial world come to be framed and remade through the lens of climate change whilst others do not is critical to any analysis of the ways in which the governing of climate change is accomplished. Alongside constituting the authorisation to improve climate change (see Chapter 3), intervention requires its own forms of delimitation. For many theorists of the policy process, such demarcations are drawn geographically (global, national, local) or in relation to specific solutions that are to be developed and implemented, whether these be particular plans or policy instruments. Yet 'what gets overlooked ... are the potentially conflicting ways of representing the environment, and the significant political work involved in settling precisely what constitutes the object to be governed' (Rutland & Aylett 2008: 628). It is precisely the nature of what it is that should be governed and how this come to be configured and contested that is central to the analysis developed here. Rather than taking the basis for intervention for granted, governing is an inherently '*problematizing* activity ... linked to the problems around which it circulates, the failings it seeks to rectify, the ills it seeks to cure' (Rose & Miller 1992: 181, emphasis in the original). Such a perspective draws our attention towards the ways in which climate change has come to be understood as a problem requiring intervention. It also suggests that the shift from the broad will to improve (Li 2007a) or govern climate change to specific programmes of intervention is neither straightforward nor guaranteed but requires ongoing work to keep the climate problem coherent and legible. This in turn will be subject to continual forms of reconfiguration, reversal and contestation. It is to this work of (re)ordering climate change that this chapter turns.

The first part of the chapter considers the ways in which the will to improve climate change comes to be formed into specific programmes of intervention. It finds with Li (2007a) that forging alignments is central to the assemblage of specific domains as requiring intervention in climate change terms and explores the practices of bricolage through which such alignments are accomplished and challenged. Through analysing the emergence of the Hexham Hydro project and the business of being low carbon at Tesco, this part of the section argues that central to the pursuit of climate government has been the constitution of assemblages through which climate change can come to be regarded as an opportunity. The second part of the chapter examines more closely how climate change comes to be known and recalibrated through the work of such interventions. Whilst this analysis, centred on the cases of the London Low Carbon Zones and the HSBC Climate Champions scheme, finds that the techniques of visualisation commonly associated with government are at work, it argues that they take place as part of a broader suite of mechanisms designed to enhance both the intelligibility and the sensibility of the assemblages through which governing climate change is ordered.

Assembling Intervention

Whilst for many of its users and critics, the concept of governmentality has become tied to an imaginary in which centres of power or individual agents seek to govern at a distance over the social body, other interpretations of Foucault's scholarship suggest that government is achieved through a heterogeneous and fragile set of practices and techniques worked through distinct sociomaterial configurations (Chapter 1; Bulkeley et al. 2015; Li 2007a; Stripple & Bulkeley 2014; Walters 2012). Central to this reading of power and the art of government is the sense that 'rather than emerging fully formed from a single source, many improvement schemes are formed through an assemblage of objectives, knowledges, techniques and practices of diverse provenance' (Li 2005: 386). Far from being a coherent or singular project, the interventions through which governmental programmes as construed are in this reading seen to be assembled 'from an existing repertoire, a matter of habit, accretion, and bricolage' (Li 2007b: 265). In their study of financial innovation, McKenzie and Pardo-Guerra (2014: 157) suggest that 'successful innovation is nearly always bricolage: the creative, ad hoc re-use of existing resources (ideas and other cultural resources as well as artefacts), not the mechanical implementation of a grand plan nor simply logical deduction from existing scientific theory'. Likewise, in the design of physical environments to create new forms of urban experience in Singapore, Lee (2014: 149–50) finds that 'techniques and rationalities are seldom new. Rather, new objects are created out of the combinations of old elements, creating a chain of equivalences that compound different logics upon each other'. In other cases, the work of assembling governmental programmes involves the recasting of existing 'solutions' in relation to new problems. Intriguingly, several examples can be found in relation to the governing of climate change. Lovell (2004: 42) suggests that along with their charismatic characteristics, the legitimacy of the nongovernmental actors involved and the credibility that could be gained from demonstrating that a policy could work in practice, the 'material presence of . . . sustainable housing developments made them liable to "capture" and rebranding' in relation to the development of the problem of low carbon housing in the United Kingdom. In Portland, Oregon, Rutland and Aylett (2008) demonstrate the work involved in remaking issues of energy efficiency into a problematic to which climate change could be attached and through which new interventions could then be marshalled, in turn recomposing urban energy and environmental governance.

Such examples suggest that understanding the ways in which governance is accomplished requires an engagement with how a range of elements and entities are mobilised in relation to one another. These forms of bricolage are not only emergent but take place in relation both to generalised logics, such as neoliberalism; the will to improve particular conditions of the population, in this case climate change; and the multiple 'specific finalities' (Foucault 2009: 99) enrolled in particular programmes, for example reducing emissions and saving money. In this sense, climate change is not a new problem, but rather one that is assembled through existing repertoires and entities. It is therefore not surprising that interventions carry the traces of other logics and practices. Moreover, attending to the practices of bricolage and assemblage involved in the making of governmental interventions 'flags agency, the hard work required to draw heterogeneous elements together, forge connections between them and sustain these connections in the face of tension. It invites analysis of how the elements of an assemblage might (or might not) be made to cohere' (Li 2007a: 264). The point of understanding governmental programmes and the interventions through which they are realised in such terms is not to 'break these down in order to identify the determining element but, rather, to trace the processes' through which they are formed (Finlayson 2011: 550). For Li (2007b), this in turn requires that we move beyond the rhetorical invocation of the pulled-together and contingent nature of governing and adopt an explicit analytic approach through which the processes of assembling interventions can be explicated. Such an analytic, she insists, must move beyond the concerns with problematisation – 'how problems come to be defined as problems in relation to particular schemes of thought, diagnoses of deficiency and promises of improvement' (Li 2007b: 265) – and attend to the multitude of practices through which governmental programmes are assembled. In her analysis of the government of community forestry, she provisionally identifies six practices that serve to constitute governmental programmes: 'forging alignment; rendering a field of intervention technical; authorizing knowledge; managing failures and contradictions; anti-politics, through which dissent is contained; and re-assemblage' (Li 2007a: 7; Li 2007b: 265; see also Bulkeley et al. 2015). Rather than viewing such practices as taking place in sequence, they can be regarded as intertwined and simultaneous such that, for example, the forging of alignments between parties to the assemblage requires the deployment of techniques that render technical the field of intervention and seek to contain its politics.

Delineating what it is that should be improved by the work of government is therefore both a complex and a practical matter through which what constitutes climate change comes to be composed and ordered. This is a matter of delineating what it is that constitutes the terrain of climate government but also of ordering the entities of which it is composed in such a way so as to create particular alignments and conjunctures through which interventions can take place. Such practices are essential to the workings of governmental programmes, for the 'relation between political rationalities and such programmes is not one of derivation or determination but of translation – both a movement from one space to another, and the expression of a particular concern in another modality' (Rose & Miller 1992: 181). Drawing on the work of Callon and Latour, Miller and Rose (1990: 9–10) argue that such forms of translation take place through 'a delicate affiliation of a loose assemblage of agents and agencies into a functioning network' such that actors come to convince one another that 'each can solve their difficulties or achieve their ends by joining forces or working along the same lines' through 'the process that Callon and Latour name as "interessement"' – 'the construction of allied interests through persuasion, intrigue, calculation or rhetoric' (Miller & Rose 1990: 10) – that serves to bound actors together in certain constellations around particular interventions. Whilst Li (2007b) rejects the term translation as conveying too unidirectional form of movement from one entity to the other, she equally stresses the importance of the ways and means through which alignments are forged in and through the practice of assemblage as central to the work of accomplishing governance. This resonates with those within science and technology studies who insist that the ways in which constituent elements of such assemblages are bought into alignment serve to order the field of intervention in relation to particular problematics (Blok 2014). Whilst scholars in the governmentality tradition have tended to focus on the importance of actors and discourse in such practices, the roots of the concept of translation/alignment within the field of actor–network studies points to the significance of including materials, artefacts, infrastructures and devices as equally central to the practice and work of alignment within any such analysis (Bulkeley et al. 2015). The cases that follow illustrate the ways in which climate change comes to be problematised and ordered through practices of bricolage that serve to draw into alignment a host of sociomaterial entities and in turn realign existing assemblages – communities, rivers, businesses – around climate change.

Summoning Micro-Hydro in Hexham

In the autumn of 2010, the Hexham Community Partnership, the Transition Tynedale Energy Group and the Hexham Town Council established a working group to explore the potential for a community-based energy scheme in the town. Initiated in 2008, Transition Tynedale is one of a number of Transition Town initiatives that have emerged in the United Kingdom since the formation of the first group in Totnes, Devon, in 2007. At the heart of the Transition Town movement has been a concern with addressing the twin issues of reducing greenhouse gas emissions in order to combat climate change and minimising dependency on fossil fuels in order to respond to the emergence of peak oil through the development of new forms of local energy and food economies (Mason & Whitehead 2012; North & Longhurst 2013). In Hexham, Transition Tynedale sought to mobilise partners around these concerns and to align these with long-standing concerns amongst policy makers and communities for energy poverty and with the growing interest in the renewable energy resources in the region. Yet despite the initial formation of the alliance between these three partners and their shared interests in addressing the problem of climate change, what might constitute an appropriate form of intervention was initially inchoate. Rather than springing fully formed from the intention to address climate change, micro-hydro had to be summoned: assembled through the articulation of various entities and intentions that had to be aligned into a coherent ordering of the space of intervention. Rather than being any kind of problem, climate change emerged in the relation amongst different concerns for energy – as an issue of security, economy and environment. In this climate-as-energy problematic, wind power generation was the first option considered, as this was favoured by one of the Transition Tynedale members:

> I've been a member of Transition Tynedale for three to four years and one of the things we've been doing is looking at alternative sources of energy and ways of getting community energy, well, initially community wind, and we actually thought we'd got a wind project but unfortunately for various reasons it ran into opposition and we had to drop it (Hexham River Hydro Interview A 2012).

Despite initial interest, wind power was seen to be limited not only by the planning objections that arose around this specific scheme but also because of the extent to which it could create either an appropriate means of financing or an adequate community-based response to the climate-as-energy problematic. As one interviewee explained:

> . . . the community aspect is really important and in that I mean one of our leading lights [in Hexham River Hydro] is a very great fan of wind power but he strongly criticises the current method under which the developer takes all the profits and the people who are in the surrounding area get nothing at all (Hexham River Hydro Interview A 2012).

In Hexham, establishing a low carbon form of community-owned local energy generation was therefore not sufficient to hold the assemblage in line. The specific sociomaterial configurations of different energy resources also came to matter in how the will to improve climate change could be aligned with other imperatives to build community and to create new forms of revenue. In a time of local authority cutbacks, the introduction by the United Kingdom government of schemes to support feed-in-tariffs led others in the partnership to see a potential run of river micro-hydro scheme 'from a different perspective – as a funding source for Hexham Community Partnership' (Hexham River Hydro Interview A 2012). The design of the scheme in this context appeared promising:

> For the first 20 years of the scheme we expect to receive the appropriate feed-in-tariff for a hydro scheme of this size. For the remainder of the life of the scheme we expect to receive income from the sale of the electricity we generate.

We expect the scheme to generate an annual surplus of *at least* £30,000 per annum, to be spent on community projects in and around Hexham (Hexham Community Partnership 2011: 4, emphasis in the original).

The gradual alignment of climate change, Transition Towns, community benefits, feed-in tariffs, the need to refurbish existing dwellings, fuel poverty and the local river was narrated through three associations, each of which saw themselves as representing different communities and individuals with professional expertise in environmental and engineering fields. Through this bricolage of diverse sociomaterial entities, the assemblage through which climate could both be governed and enable the improvement of other conditions took shape. Momentum began to build around the development of a project to build a 100KW community hydro-power generation scheme using the modern version of the Archimedes Screw. The significant up-front costs involved in the development of both the feasibility studies for such a scheme and the capital outlay required for its investment might have proved too much for this 'delicate affiliation' were it not for the coincidental emergence of the Energyshare scheme. Initially launched as a partnership between River Cottage (established by the chef, food writer and campaigner Hugh Fearnley Whittingstall in 1998) and the energy retailer British Gas, the Energyshare scheme was launched in May 2011, and its first phase ran until the end of 2012 (see Chapter 2). Both parties saw the scheme as a means of realising their ambitions in relation to community-based energy. Energyshare is described by River Cottage as a 'renewable energy community' that aims to:

... give everyone the opportunity to source, use and (when you're ready...) even generate your own 100% British renewable energy – and save money doing it and to bring together people, communities and organisations to turn the system on its head (River Cottage 2013).

For British Gas, working more closely with communities was seen as an opportunity both to enhance their customer base and to consider new forms of business operation that tapped into the emerging community engagement with renewable energy generation:

I think probably at the beginning of that relationship we were looking at partners where we could probably sell more gas and electricity, and Energyshare certainly for us it was very much about tariff and but there were some great byproducts that came from that in terms of, you know, access to communities because we knew that there was a big community hotbed out there (British Gas Interview 2012).

At the heart of this phase of the Energyshare scheme was a competition for community-based energy schemes to win £100,000 in each of three categories based on the size of the project they wished to develop (see Chapter 2). Having conceived the notion of a micro-hydro scheme, in May 2011 Hexham River Hydro (HRH) applied to the competition in the large project category, were shortlisted in November 2011 and gained enough public online votes to win the competition in December 2011:

> There was a lot of kind of nervous energy because we did recognise we were only a small town, you know Hexham is just over eleven thousand [people], and although it's a regional thing, you know, the kind of support for it wanes a little as you get away from the actual town, we were competing with bigger places who have got access to bigger populations and it's all online and it is you know sudden death, you've either done it or you haven't on noon on a Saturday, wasn't it, high noon on a Saturday was the cutoff so you don't know until the end what the result is going to be (Hexham River Hydro Interview B 2012).

Entering the Energyshare competition required that Hexham River Hydro undertake ongoing work to translate hydro power through the local community and the transnational Transition Towns network. Familiar techniques of public outreach and symbols of the importance of local responses to climate change were mobilised as devices to achieve this effect, including media opportunities and photo calls (involving local Members of Parliament, the MEP, councillors and representatives from community organisations in Hexham), which used an inflatable globe with the message 'Support HRH – Save Tomorrow Today'. There was an interview on Radio Newcastle to raise the profile regionally, reports and photographs in local and regional press (Journal 2011a and 2011b), discussions with local schools, attendance at workshops on green issues to engage a range of private, public and voluntary groups, and e-mails to other Transition Town groups globally (whose members were, perhaps curiously, also allowed to vote online for the HRH scheme) (Energyshare 2011). In this work, climate change remained a means through which diverse interests could be orchestrated, and the authorisation of the scheme could be achieved by emphasising the exemplary nature of the hydro scheme and the ways in which it was part of a bigger project (see Chapter 3):

> It is now widely accepted that the empirical evidence of increasing global temperatures is the cumulative effect of anthropogenic greenhouse gas emissions

(GHG). There is currently a global challenge to reduce GHG to mitigate against the effects of climate change and limit the temperature rise to 2°C, in line with the Copenhagen Accord goal. In 2008 the UK Climate Change Act set legally binding targets for the UK to reduce greenhouse gas emissions by at least 80% by 2050, set against a 1990 baseline. To assist these reduction targets the Government has pledged to generate 15% of all energy consumed, by 2020, from renewable sources. Renewables currently supply only 6.7% of total UK electricity demand (Hexham Community Partnership 2011: 2).

Yet following the award of the Energyshare funding, as the Hexham River Hydro project came to be reworked through the development of the feasibility study and various forms of financial and environmental assessment, climate change began to disappear from view:

> ... it's definitely not about climate change, it's just that climate change is a difficult concept, so is sustainable development but that's so woolly you can actually explain the story a bit better (Hexham River Hydro Interview A 2012).

The overall contribution of the scheme to reducing greenhouse gas emissions purely through the generation of energy was small, and its wider impact depended on the generation of streams of revenue that could then be deployed in other regeneration and energy savings schemes. As financial uncertainties around the feed-in-tariff, coupled with growing levels of resistance from publics whose constitution was being challenged by the presence of hydro power in the river, placed the scheme at risk, the associational power of climate change as a means to govern the assemblage began to wane and was overtly questioned by those who felt that such a scheme threatened the resilience of the river in the face of the need for ecological adaptation (Chapter 6; Armstrong & Bulkeley 2014). These refractory processes, excluded from the initial assemblage of the government of climate change through community energy, serve continually to contest and disrupt the fragile alignment through which intervention is assembled (Li 2007b). Here the work of alignment is intimately related to the practices of problematisation, through which the multiple problematics, and opportunities, of relating sociomaterial conditions to climate change come under question.

Building Climate Business at Tesco

> A low-carbon strategy is also vital if we are to minimize the risk to our business: the physical threat of climate damage to

> our supply chains, the resulting economic damage; and the serious effects of rushed and inefficient regulation if we fail to act in time and governments are forced to take draconian action (Leahy 2009).

> But this is about more than just mitigating risk. For Tesco a revolution in green consumption is a fantastic opportunity: once and for all to break the link between consumption and emissions, and in doing so to satisfy a new consumer need, and grow our business. That is the goal of a sustainable business (Leahy 2009).

In the 2000s, climate change came to be seen as a problem by and for United Kingdom corporations (see Chapter 2). If during the 1990s a range of fossil fuel interests had sought to deny or derail the governing of climate change, by the mid-2000s a growing number of corporations had come to see climate change as both a risk and opportunity for their business, as the extracts from then–Tesco Chief Executive Sir Terry Leahy suggest, and to regard themselves as part of the means through which the improvement of the condition of climate change should be governed. Tesco, at the time the third largest retailer in the world, was one of the first supermarkets to align itself with the climate change agenda (Gouldson & Sullivan 2012). Tesco's engagement with climate change started in 2007, when Leahy pledged to reduce carbon emissions of the business by 50 per cent by 2020. In a follow up speech in 2009, he further committed Tesco to become a zero carbon business by 2050, a goal still retained by the corporation as one of the 'essential' aspects of their corporate sustainability (Tesco 2014). Central to the work of governing climate change within Tesco, of establishing the basis for intervention and the authorisation to do so, was the casting of the problem as one of the need for leadership and collective action beyond the nation-state:

> . . . one of the motivators for taking a leadership position is because we're not seeing leadership coming from anyone else at this point. So, you know, we're pushing ahead and our targets are more challenging than the other ones and we're hoping that their ideas change as well (Tesco Interview A 2011).

> . . . of course, you want to work with international frameworks and government frameworks; but I certainly think that the failure – or perceived failure of Copenhagen – and the agreements reached in Cancun . . . they galvanise us

> further. They serve to make us feel like we're going in the right direction, you know, we can actually do this. We're in a privileged position of not having to negotiate around that sort of table (Tesco Interview A 2011).

This positioning of Tesco as an entity that could govern climate change beyond the state was achieved through its alignment with the building of a consensus, which gathered pace during the 2000s, that the issue was a problem requiring collective action by business and government (see Chapter 2). Tesco's chief executive and its board of directors were regarded as a crucial node in this consensus, creating a 'sense of responsibility', which was translated into 'a leadership position' that 'for an organisation like ours ... has to come from that level of personal commitment' for the need for a corporate response (Interviewee Tesco Head Office). To this end, Tesco also became involved in the work of the Climate Group. Formed in 2004, this nongovernmental organisation, in its words, 'brought together a handful of leaders and innovators from governments and businesses around the world' a 'group of pioneers' who 'were the first to see the positive business case for climate action' (Climate Group 2014d). Tesco became a high-profile member of the activities of the Climate Group, including the 'We're in This Together' campaign through which then–United Kingdom prime minister Tony Blair sought to engage leading United Kingdom businesses in bridging the trans-Atlantic divide on climate politics whilst also seeking to shore up public opinion and action (Climate Group 2007).

In seeking to translate climate change as a corporate problem, Tesco set a series of targets aimed at its own operations, at the emissions accruing through the corporation's supply chain and through the actions of their customers, pledging to reduce 'emissions of the products in our supply chain by 30% by 2020' and to find 'ways to help our customers reduce their own carbon footprints by 50% by 2020' (Tesco 2010: 8). Rather than being confined to an external threat, a matter of efficiency or one of corporate social responsibility narrowly defined, the problem of climate change was expansive within Tesco, coming to figure in relation to the ways in which it conducted itself both within and beyond the boundaries of the corporation. Climate change, and in particular carbon (see Chapter 5), came to be a way in which Tesco started to view its business in different terms as new sites and processes entered into the field as requiring intervention, for example in terms of the corporation's property portfolio:

> A lot of it is in property. And the thing that surprised me when I looked at the numbers is how much of it is in refrigerant leaks, actually. ... So electricity and then other things like [refrigerant] gas are a big part of our [carbon] footprint . . . and some of the things that we do in the existing stores and then also in the new ones is energy efficient lighting, doors on freezers and all sorts of things that you think about (Tesco Interview A 2011).

Yet whilst central to the constitution of the climate problematic as a corporate issue, such forms of internal carbon management were not seen as sufficient:

> If we're going to get down to the targets we've set ourselves for 2020, but equally, also for 2050, it can't just be about energy efficiency. Because you still have to have some energy to run these things. And so a big challenge for us is about how we develop renewables that help us do that ... we've got a certain amount of things you can do on-site, in zero carbon stores we've got biofuel and CHPs [Combined Heat and Power], and we've done a bit of wind and we're looking at some other things, PV, that kind of thing, and you can do quite a bit, because you've got quite a lot of stores and you've got land associated with that ... but get further than that and, well, how do you do [generate renewable energy] . . . off-site? And what does that mean for the business name and corporation? (Tesco Interview A 2011).

The problem of climate change was not simply then one of managing emissions, but of thinking about a different way of being a business – one that, for example, might produce its own renewable energy or that organised its supply chains in different ways to reduce carbon footprints. The business case for building climate change into the body corporate in this way rested both on costs and the potential savings that could ensue from efficiency measures and securing renewable supplies in a context of the energy price rises that dominated the late 2000s and early 2010s, but also in terms of innovation and the creation of new technologies that could sustain energy and cost savings over the long term. The logic of building the value of carbon saving through the business was also extended to consumers:

> . . . we actually know what thirty positive, relatively low-cost actions are, that we all could do today – and that would halve our carbon footprint today. So it's not, you know, halving our customer's footprint for 2020 isn't, by any stretch of the

imagination, outlandish. But to get people to understand that it isn't about not trying to be awkward, and not taking that long, well-deserved holiday you've been saving up for, but about ventilation or using a decent boiler (Tesco Interview A 2011).

Seeking to address climate change as a corporate concern through Tesco involved the alignment of a host of sociomaterial entities, from boilers to refrigerants, stores to homes, money to carbon, which in turn required translation through different forms of business practice. The rendition of the will to improve climate change into a matter of government, manifest through multiple schemes and techniques, required the ongoing work of bricolage, of interweaving corporate practices of social responsibility, partnership, commerce, investment with one another and in relation to the opportunities of being in the climate business and the political force being generated through the formation of a coalition of state-based and business interests. Existing practices, of managing the finances of the company and the budgets of consumers, were redirected towards new ends whilst new techniques, including the assessment of carbon neutrality, were developed. The assemblage ranged far and wide, through the Climate Group, and across the multiplicity of sites and employees incorporated in the business of Tesco. In contrast to the case of Hexham River Hydro, where the governing of climate change summoned a new entity – a micro-hydro installation – which gathered momentum and opposition, and whose insertion into the landscape attracted outright opposition (see Chapter 6), at Tesco contestation over what it meant to engage a corporate response to climate change emerged in the techniques of calculation and commensuration and through which the practice of calculating climate change itself took place (see Chapter 5). Yet in both cases, the work of assembling governmental programmes depended not so much on forging alignments around 'diagnoses of deficiency' (Li 2007b: 265) or problematisation in the most straightforward sense of identifying specific problems or conditions that required correction, but rather on what might be termed *diagnoses of possibility*, of the opportunities that climate change offered for improvement, whether that be in relation to securing local energy resources, creating new business opportunities, providing leadership and so on. In each case, the disposition of government is not concerned with singular problems or threats, but rather such governmental programmes encompass 'combinative strategies that seek to preempt, prepare, and offer protection against a range of threats, determinate and indeterminate, banal and extraordinary, as well as a range of aspirations' (Kraftl 2014: 282). Rather than being only a matter of rendering climate change technical, each case also illuminates the important ways in which making climate change compelling or desirable is also central to accomplishing governance (Bulkeley et al. 2015; Deschenu & Paterson 2011; Paterson & Stripple 2012).

Comprehending Climate Change

As the preceding cases suggest, at work in the ordering of climate change from a generalisable claim for improvement to a particular kind of problem requiring specific forms of intervention at different sites and moments are the interwoven practices of alignment and problematisation through which governmental programmes are assembled. In addition to the work of bricolage, translation and interessment involved in composing interventions, such work requires that an arena of intervention is rendered technical, that is that it comes to be represented as 'an intelligible field with specifiable limits and particular characteristics ... defining boundaries, rendering that within them visible, assembling information about that which is included and devising techniques to mobilise the forces and entities thus revealed' (Rose 1999: 33; quoted in Li 2007a: 7). Accomplishing governance, in short, requires that the arena of intervention is *comprehended*, where to comprehend includes 'to seize, grasp, lay hold of, catch. ... To take in, comprise, include, contain. ... to lay hold of all the points of (any thing) and include them within the compass of a description or expression; to embrace' (*Oxford English Dictionary* 2015). Constituting a field or domain as viable for intervention is of course not separate from the practices of problematisation through which the nature of the improvement required and how it is constituted is conceived, for problems and their solutions are mutually configured (Li 2007a: 7; see also Hajer 1995; Lovell 2004). Further, the drawing of field boundaries also serves to exclude other potential problematisations, unruly elements, refractory practices and forms of resistance (Li 2007b). Contestation around these boundaries serves as a space where the political moments of governing climate change gather (see Chapter 6).

The means through which governmental programmes are rendered technical are therefore significant in relation to the ways in which the intention to govern climate change is translated and enacted. Whilst Li

(2007a: 7) suggests that a whole host of practices and techniques are involved in the process of rendering technical, analysis has tended to focus on just a few categories. Much attention has been given to the ways in which such fields of intervention come to be circumscribed and governmental programmes pursued through the development of (state-based) science, statistics and other forms of calculation. In their account of the ways in which government comes to be conducted, for example, Miller and Rose (1990: 5) single out such calculative practices as central:

> 'Knowing' an object in such a way that it can be governed is more than a purely speculative activity: it requires the invention of procedures of notation, ways of collecting and presenting statistics, the transportation of these to centres where calculations and judgements can be made and so forth. It is through such procedures of inscription that the diverse domains of 'governmentality' are made up, that 'objects' such as the economy, the enterprise, the social field and the family are rendered in a particular conceptual form and made amenable to intervention and regulation.

Such practices are also central to Scott's (1998) argument concerning how state power is manifest through techniques that enable particular domains to be made legible, simplified in order to conform to the ideals and intentions of centralised powers (Lee 2014). Legibility, whether through visualisation or other forms of representing society, is not a given but rather 'a condition of manipulation' (Scott 1998: 183). Reading such techniques through a governmentality lens, Li (2005) argues that whilst important in attending to the practice and politics through which state power is manifest, Scott's account overplays the extent to which such techniques simplify the sociomaterial world, are able to bring it to order or exclude other situated ways of knowing. Yet there remains a tendency for scholars of governmentality, particularly those concerned with neoliberal forms of power, to remain focused on the ways in which fields of intervention are comprised and represented through techniques based on rationality and calculation in the relatively narrow sense of these terms. This in turn has fuelled analysis of the ways in which government can be exercised 'at a distance' over time and space through the deployment of relatively immutable devices that conduct governmental intent, serving to reinforce the 'programmers view' of governmentality (Death 2014) and negating the more proximate and practical means through which governing is assembled.

Whilst such techniques may be critical means 'through which governmental programmes construct intelligible fields for intervention' (Ghertner

2010: 186), this narrowing of vision serves to limit the ways in which we can understand how such fields come to be composed and delimited. Recent scholarship suggests that such analysis can be extended in two important directions. First, there is considerable scope to move beyond notions of calculation in a strict, scientific sense in order to engage with the ways in which 'governmentality can operate as effectively through aesthetic norms as it does through those "scientifically rational" and statistical processes of knowledge assembly widely discussed in the literature' (Ghertner 2010: 186). For example, through his analysis of the government of slum areas in Delhi, Ghetner (2010) demonstrates how the failures of calculative techniques served to reconfigure what constituted the domain for intervention in relation to slum clearance through aesthetics of visibility and cleanliness. What was calculated here was not only the matter of rational assessment of the state of cleanliness of any particular street but of what the proper aesthetic of being clean entailed.

Second, there is significant potential to acknowledge the ways in which the legibility of any domain to be governed is constituted not only through discourse or visualisation, forms of relating that are in some sense extractive, but through various forms of experience and encounter that are more immersive. In other words, coming to know a problematic as in need of intervention and to circumscribe its limits entails active and practical forms of involvement with its specificities. For Lee (2014), the question is not then one of 'seeing' like a state, but rather of 'feeling' like a state, as forms of governmental intervention come to be orchestrated by the sensory experience of the city. Rather than conceiving of the practice of government as concerned only with the *intelligibility* of intervention, a term that draws attention to the intellectual means through which understanding is generated, such approaches also suggest we need to attend to the ways in which governing is accomplished through configuring and being configured by the *sensibility* of intervention – that is, the 'power or faculty of feeling, capacity of sensation and emotion as distinguished from cognition and will' (*Oxford English Dictionary* 2014) by which governmental programmes also come to be known, delimited and narrated. Such techniques and practices exceed the rendering technical of any particular domain. Instead, orchestrating the will to improve requires a multiplicity of practices that serve to connect and constitute the assemblage through *rendering comprehensible* the field of intervention and the governmental programmes through which it is circumscribed and configured. The cases of the Low Carbon Zone in Brixton, London, and the development of the Climate Champions scheme within HSBC illustrate the ways in which intelligibility and sensibility combine in order that climate change can be comprehended and acted upon.

Delimiting London's Low Carbon Zones

During the 2000s, the first elected mayor of the newly constituted Greater London Authority, Ken Livingstone, established climate change as an issue requiring intervention within the city (see Chapter 2). Over the course of his two terms of office (2000–8), Livingstone orchestrated the foundation of the C40 Cities Climate Leadership Group, established the London Climate Change Agency as an arm's-length public–private organisation, developed the 2007 Climate Change Strategy that adopted ambitious long-term targets for the reduction of greenhouse gas emissions and the generation of decentralised energy and promoted a range of schemes aimed at fostering energy efficiency improvements in the domestic and commercial built environment (Bulkeley & Schroeder 2008). The consensus that London should engage in the governing of climate change, and moreover that climate governance could improve London, was forged across public, private and party political divides, such that when the Conservative and somewhat sceptical Mayor Boris Johnson was elected in 2008, the issue remained one around which governmental intervention was orchestrated (Bulkeley & Schroeder 2012; Bulkeley et al. 2014). Seeking to renew this commitment and demonstrate the possibilities for realising urban responses to climate change, the Low Carbon Zone (LCZ) programme (initially announced in July 2008 and formally launched in May 2009) designated ten areas across London as test-beds for community-driven approaches that could meet a target of bringing about a 20.12 per cent reduction in carbon emissions by 2012 (in time for the London Olympics) and to put in place mechanisms to achieve a further 60 per cent carbon reduction by 2025 (in line with the targets contained within the 2007 Climate Action Plan).

At first glance, the demarcation of ten LCZs across the map of Greater London conveys the all-seeing power of the state and its attempts to simplify what it means to govern carbon. As Scott (1998: 54–5) argues, 'stopping short of redesigning cities in order to make them more legible ... state authorities endeavoured to map complex, old cities in a way that would facilitate policing and control.' What is more, the 'grandiose, high modern, state-driven' projects with which much of Scott's analysis is concerned 'often retreated into miniaturism . . . micro-managing confined spaces and producing visual effects that were all the more striking because of their spatial concentration' (Li 2005: 386). Similarly, rendering the field of intervention for climate government comprehensible through the LCZ programme served as a means through which the will to improve climate change could be readily contained, with local authorities asked to delimit areas of up to one thousand residential, commercial and public sector buildings and to bid for seed funding up to £400,000 in order to facilitate intensive carbon reduction activities. Yet rather than seeking to simplify or rationalise urban space through this means, the LCZs were constituted as dynamic, complex and unfolding spaces through which experimentation and demonstration would serve to produce new ways of knowing how to govern carbon:

> These energy busting zones will create an armada of flagships across London, focused on finding the most effective ways to rapidly cut carbon and slash energy bills (Greater London Authority 2009a).

> A Low Carbon Zone (LCZ) is a local geographic area with concentrated activity to try out innovative ways of reducing CO2 emissions. The LCZ programme will bring together local authorities, business and industry, public sector and community organisations to develop, demonstrate and measure the impact of initiatives that create a big reduction in CO2 emissions. . . . LCZs will act as exemplars for London, showcasing technological solutions and catalysing long-term reductions in CO2 more broadly across the capital (Greater London Authority 2009b: 29–30).

> Although each zone contains a mix of residential, community and commercial buildings, the zones vary in size, location and building construction. Some are urban and high density, some are town centres and others are suburban residential areas. Some zones are generally affluent; others are areas of high deprivation. Some contain communities that have a high level of social capital; others are more disengaged. This mix was selected to reflect the diversity of London and allows an understanding to develop of the types of approaches to carbon saving and local quality of life that are appropriate in different circumstances. This learning will be crucial to the development of London as a leading low carbon city (Greater London Authority 2011a).

Central to the constitution of LCZ as a means through which climate change, and carbon specifically, can be governed is an engagement with the multiplicity and materiality of the city. In its successful bid to host an LCZ in Brixton, South London, the Lambeth Borough Council (one of thirty-two borough councils that make up the area of Greater London) explicitly sought to align itself with the ambitions of the mayor and the GLA in relation to innovation and community building:

> Brixton is a vibrant hub of carbon reduction activity and innovation with the Future Brixton Masterplan aspiring to One Planet Living status, Brixton Green's zero carbon development proposals and Transition Town Brixton (London Borough of Lambeth 2009a: 1).

Following their successful bid, Brixton LCZ was launched in March 2010 and contains around 3,500 properties, including ten high-rise and thirty-six low-rise blocks, street properties (social and private housing) and commercial and public sector buildings, with the aim to 'get every business and household within Brixton Low Carbon Zone involved in carbon reduction' (London Borough of Lambeth 2010). Much of the zone falls within one of the most deprived wards in England, with high unemployment, high levels of fuel poverty and the majority of properties consisting of social housing. Rather than seeking to place these potentially problematic conditions outside of the LCZ, they were actively incorporated:

> Brixton was chosen as the most suitable area of Lambeth for a Low Carbon Zone by checking all potential areas in the borough against the Low Carbon Zone criteria. Brixton has the most suitable range of buildings and existing projects. Brixton also has high levels of deprivation, fuel poverty and a significant need for carbon reduction measures. In addition, Brixton has the potential to attract funding and there is scope to build on existing community engagement initiatives (London Borough of Lambeth 2010).

This was, as the quote suggests, on the one hand a process of rational calculated activity, of checking criteria and fit; yet on the other hand, the designation of the LCZ depended on sensing the potential and appetite for low carbon in Brixton, the commitment and values of partners and the sorts of opportunities that could be leveraged through purposefully drawing the boundaries of the zone to encompass those areas of the local authority that were eligible for improvement by other means, including the Community Energy Saving Programme (CESP) retrofitting programme and the refurbishment finance of the local authority:

> I feel they were fairly confident at the time that they were putting in the bid that all of these other programmes would take place, like CESP would take place, various other inputs ... in terms of boiler replacements, and some roof insulation work. So in terms of residential properties, they were fairly confident that there would,

through Lambeth's capital allocation programme, there would be work happening within the zone. So I suppose that element was kind of a bit sorted. So they used the GLA money to bring in the added value (Brixton Interview A 2011).

... the reason why Brixton was chosen was because there was a lot of energy and projects already happening, in the pipeline to happen, there was a big piece of work ... Future Brixton, and it's a master plan done with heavy consultation with residents and transition towns externally (Brixton Interview A 2011).

Rather than serving to confine the field of intervention, the bounding of the LCZ in Brixton provided a means through which to comprehend low carbon as a means through which multiple 'specific finalities' (Foucault 2009: 99) could be improved in conjunction with one another. This has required practices through which the Zone has become legible, demarcating the specific buildings and residents who are eligible for financial assistance, but it has also involved marshalling wider networks through the LCZ, such as Transition Towns Brixton, and the formation of a number of schemes – including Green Doctors, Draughbusters and Brixton Energy – whose reach extends beyond the geographical limits of the LCZ. At the same time, making Brixton LCZ required marshalling the 'sensibility' of the sociomaterial configurations at hand, through convening the partners, drawing in the unruly and unimproved buildings and building on different forms of community engagement and 'know-how'. Through these practices of assemblage, rather than serving to simplify or demarcate the space of climate governance, Brixton LCZ enabled the establishment of the field of intervention for governing climate change as one that needed to cut across the traditionally separate domains of energy, environment and housing and to address matters of energy security as well as carbon reduction. This has served as a means through which different forms of capital can be assembled and drawn together on the one hand, but also to enable the formation of new climate publics through which what it means to respond to climate change is articulated and contested (see Chapter 6).

Configuring Climate Champions

> ... there are particular physical qualities of the atmosphere that, in marked contrast to other forms of environment (including land, forests and hydrological systems), makes its government particularly dependent upon the technologies of the imagination (Whitehead 2009: 220-1).

Central to integrating the governing of climate change into the body corporate at HSBC was the formation of the Climate Partnership, a five-year programme (2007–11) led by HSBC, with the Climate Group, Earthwatch, WWF and the Smithsonian Tropical Research Institute (see Chapter 3). Within the Partnership, which included interventions to foster urban responses to climate change in the major global cities in which HSBC is based and conservation of vulnerable ecosystems, a critical component was the development of the Climate Champions scheme. Designed by Rachel Phillips, the head of learning at Earthwatch, the overall aim of scheme was to 'develop environmental leaders (climate champions) amongst HSBC employees and embed sustainability into the DNA of their business' (Earthwatch 2013a). The scheme involved three main steps: attendance at a field course, the development of a workplace project to address climate change and participation in a broad range of forums and mechanisms to exchange experiences and learning. Participants ranged from senior management to local employees in different branches of the organisation across North America, South America, Europe, Asia and the Pacific, with the result that during the five years of the Climate Partnership Earthwatch engaged 2, 267 HSBC employees from 65 countries in the Climate Champions scheme (Earthwatch 2013b). Through this means, the notion of corporate social responsibility shifted from being a matter of acting beyond the corporation to one in which climate change was to become incorporated as a matter through which the body corporate could itself be improved, with the aim to:

> Engage thousands of people in climate action and research, equipping them to act as 'climate champions' in their workplace and community . . . [and] Build HSBC employees' capability and commitment to sustainable business practice (Climate Group 2012).

The ways in which the field of intervention came to be comprehended through the Climate Champions scheme were central to the way in which climate change came to be regarded as an issue requiring intervention by and through HSBC as well as to become the means through which this was practised across the body corporate. As Whitehead (2009) suggests, the ephemerality of the atmosphere means that unlike other environmental conditions, its government requires both the use of scientific observation and the development of 'technologies of the imagination'. Similarly, the Climate Champions scheme sought to contribute to the observation and recording of climate change whilst also working to 'increase [the] *environmental sensibility*' (HSBC Partner Interview B 2010, emphasis added) of the bank and its employees.

Drawing on their previous partnership, *Investing in Nature*, Earthwatch and HSBC sought to sustain and expand their mutual engagement in the development of field-based studies of the changing environment through the Climate Champions scheme. As Earthwatch explain on their website,

> Data validation, establishing the accuracy or bias in volunteer measurements, is a crucial part of utilising citizen science data, especially as there is growing involvement of citizen science in, for example, large-scale or long-term ecological monitoring projects. Citizen science can play a valuable part in the type of ecological monitoring that is usually costly, time consuming and labour intensive (Earthwatch 2013c).

The explicit role of Climate Champions in creating scientific assessments about the impacts of climate change on regional forest systems serves a dual purpose. On the one hand, it contributes towards the corporate social responsibility goals of HSBC to donate time and resources to deserving projects whilst also enabling it to play a part in the accumulating consensus about climate change (see Chapter 3). On the other hand, it provides a means through which employees are engaged, firsthand, with the climate problem:

> I mean, it's all very well reading about things and being taught things, but when you see things firsthand it brings it home to you, if you like. So, for example, part of the field study was measuring different types of trees; the height of them, the circumference of the trunks. It was taught, in a way that, okay, what we were recording now would in, say, four or five years' time, would be slightly different to what other groups would record. You've got subtle shifts in small mammal species or tree species which may indicate some sort of climate shift, climate change. So it . . . we were doing a fair amount of work, fieldwork, which brought everyone together (HSBC Interview B 2010).

> It kind of grounded you, if you like, in what it actually means, all this, in a way. Because, you know, in a city you can by and large ignore the weather, most of the time. Whereas when you're staying somewhere like that and doing things outside, you know, the weather is so much more important to you. And then the effects on everything in forests. So it's important, I think – especially for people like us who work in these kind of environments – it's really important to get out there and take a look at it all. Because I think it drove home to people what it meant. What it could mean (HSBC Partner Interview A 2010).

Rather than participants relying on secondhand knowledge delivered through lectures or online material, essential to the work of the Climate Champions scheme is the practice of enabling HSBC employees to encounter climate change through *bearing witness* to the changing environment. Fieldwork acted not only as a calculating technique, but also as a 'technology of the imagination' (Whitehead 2009: 221) through which individuals came to align their own conduct with the climate change problematic. As one of HSBC's partners observed, for those involved, 'it's a kind of impactful personal experience. And I guess that's what we try to do, is kind of impact them at a personal level but also, in terms of their role in the business as well' (HSBC Partner Interview B 2010). Moreover, the shared and visceral experience of field research served as a means through which Climate Champions came to configure and be configured through HSBC as a climate business:

> We were very much encouraged to go on these things and we know that it will reflect well upon us and that we can really make a bit of a difference. I think if I went on it outside of work it would be good and I'd enjoy it but it would be for self-interest more than anything. Whereas when you know you've got the weight of HSBC behind you, it's quite powerful in terms of going back to the workplace and knowing that you can really make some big changes (HSBC Interview C 2010).

> 'What does the term climate champion mean? 'Climate champion means we have to answer the call of HSBC to make HSBC the most environmentally sensitive company in the world bar none' (Participant A); 'as a climate champion I believe that I will be able to make a difference, not just within HSBC, but also in my community' (Participant B); 'I think the word climate champion for me in a nutshell means a person who is willing to take responsibility for climate issues and convey that, and act upon it, acknowledge that there is an issue, accept that there is an issue, and act' (Participant C). (Smith 2013).

The Climate Champions scheme serves as a means through which the corporate conduct of climate government comes to be shaped and circumscribed, through techniques that work to render the intervention both legible and 'sensible', subject to shared affective understanding. As Whitehead argues, such forms of demonstrable and experiential work, in his case 'exhibitions and training schemes[,] . . . reveal a more intimate expression of the power of government to shape the everyday conduct of citizens than those associated with the regulations of government at a distance' (Whitehead 2009: 223; see also Hajer 1995: 271). Enacting such conduct within the workplace proved an altogether more challenging and contested process. For the senior managers at HSBC, the Climate Champions scheme was central to what they termed 'joining up the bank' around the climate business, orchestrated through drawing people from across the business together through the fieldwork sites, an online platform and the social networks that were established by the Champions themselves. Yet despite the enthusiasm generated, once Climate Champions returned to the office and attempted to enact the projects designed to improve the corporate response to climate change, they found that the free reign they had been given to devise the 'most weird and wonderful and wacky ideas' was heavily circumscribed such that '[only] some of [the ideas] can be implemented, and most ... probably can't. And so that can be potentially demotivating' (HSBC Interview D 2010). Translating the climate problematic as comprehended through the field and in the early stages of the Climate Champions scheme within business as usual required new practices of calculation and commensuration in order to make the projects 'fit' within HSBC, a set of practices explored in greater detail in Chapter 5.

Rather than being a finished project of demarcating clear boundaries, of arenas through which intervention proceeds through practices of rational and calculated action, these cases suggest that rendering such fields comprehensible requires both ensuring their legibility and their sensibility. Such work is a 'cautious and multi-step process, incorporating varying degrees of uncertainty and reflexivity' (Lee 2014: 150) as it unfolds. The sociomaterial assemblage through and of which climate government is conducted is neither fixed nor assured, but in a constant process of being formed and recalibrated. In accomplishing climate government, 'priorities and agencies emerge ... as actants come together, and work together, to produce an order. Any order that is created, finally, is temporary, precarious; it requires ongoing work to keep it alive' (Rutland & Aylett 2008: 633). There is no one climate order, no predetermined climate to be governed, rather an ongoing (political) process through which what climate government entails is ordered, contested and reconfigured.

Conclusions

Bringing climate to order is part of the work of government. Rather than being a defined problem with clearly legible solutions, what climate change means and how it should be addressed are made through the processes of accomplishing governance. The climate problematic is composed through those interventions that seek to improve it. In parallel to practices of authorisation (see Chapter 3), the will to improve works through aligning and configuring the field or domain of intervention within which climate change can be ordered and comprehended. Such practices of composition not only require new objects – images of global temperature curves or sensory perceptions of the changing seasons, for example – but perhaps more importantly serve to charge all manner of existing objects – from the contents of a supermarket basket to the flow of a local river – as having some form of intelligibility or sensibility in relation to climate change. How energy should be provided, how customers should shop, how cities should be renovated and how banks should go about their business, to take the examples discussed in this chapter, come to be considered as matters that should be comprehended in relation to climate change. This is not a seamless endeavour, but rather a process of bricolage, of the collection of seemingly miscellaneous entities that can be made to fit into climate change and with which climate change can be forged into a problem suitable for intervention.

Yet to attend to the ways in which accomplishing climate governance happens through these practices of assemblage is not to deny its purposeful nature. Some entities are enrolled and aligned, bought into a climate order and rendered both intelligible and sensible in climate change terms. Others either struggle to be accommodated or are neglected altogether. Moreover, climate change comes to be ordered in different ways. It is at once a problem of energy security and the logistics of supply chains, of how much paper is used in the photocopiers and of the design of housing estates. As Rose and Miller (1992: 82) suggest, 'such translatability between the moralities, epistemologies and idioms of political power and the government of a specific problem space, establishes a mutuality between what is desirable and what can be made possible through the calculated activities of political forces'. The processes of alignment and comprehension through which climate change governance is accomplished serve to translate the 'improvement' of climate change through practices and techniques that are 'not just governed by institutions, prescribed by ideologies', but also 'up to a point, possess their own specific regularities, logic, strategy, self-evidence and "reason"' (Foucault 2000b 225). Each ordering, alignment and assemblage renders climate change comprehensible and governable in distinct ways, yet the 'resonance of the label itself' serves to 'flag an identifiable terrain of action and debate' through which intervention is both composed and directed (Li 2007b: 267). Rather than expecting common solutions to a common problem, such analysis suggests that accomplishing climate governance may instead entail fostering its diversity, contradictions and creativity. It is to this diverse array of practices and the forms of governing that ensue that Chapter 5 turns.

5

Climate Government Articulated

Introduction

Programmes of improvement are far from straightforward. Not only do conditions, problems and potential have to be drawn into alignment and come to be known and delimited, but ongoing work is required to achieve the 'right way of arranging (disposer) things in order to lead (conduire) . . . not to the form of the "common good" . . . but to a "suitable end", an end suitable for each of the things to be governed' (Foucault 2009: 99). This is work that takes a multitude of forms. Think for example of car driving:

> Airbags, emissions guidelines, car registration plates, road signs, advertising campaigns, tracking devices and a host of other material and immaterial things have been devised and deployed by experts to facilitate the government, movement and performance of people in cars: identifying and categorising vehicles, encouraging, maintaining or enhancing their movements, improving their comfort, security and safety, persuading them to alter their conduct, and aiding law enforcement (Merriman 2005: 235).

As Merriman's work suggests, the government of driving involves a host of material artefacts, codes and other devices aimed to arrange 'a sort of complex of men and things' (Foucault, 2009: 96) in such a manner that they 'will do as they ought' (Li 2007a: 5). Accomplishing governmental programmes depends upon 'the techniques, the practices, that give [them] concrete form' (Foucault 2000b: 410). What Callon and colleagues (2007) term devices are central to understanding how government is conducted through an assemblage or dispositif, for they help to 'account for the connections that hold highly discontinuous elements together' (Hébert 2014: 22).

Such devices, techniques and practices, as the case of car driving indicates, may be rather mundane and incremental but are nonetheless critical to the ways in which 'authorities seek to instantiate government' (Rose & Miller 1992: 183). At the heart of the analysis of the ways in which governing is conducted, therefore, has been a concern to reveal and investigate the taken-for-granted ways in which the sociomaterial world is ordered and configured through, for example, 'techniques of notation, computation and calculation; procedures of examination and assessment; the inventions of devices such as surveys and presentational forms such as tables; the standardisation of systems for training and the inculcation of habits; the inauguration of professional specialisms and vocabularies; building designs and architectural forms – the list is heterogeneous and in principle unlimited' (Rose & Miller 1992: 184). Despite this potential array of different techniques, analytic attention has been centrally concerned with the practices of calculation and standardisation (Li 2007a; Lovell & MacKenzie 2011; Rydin 2007). Such techniques and practices are critical, for they are essential to the means through which governmental power can be exercised at a distance, relayed over time and space, because of their capacity to mobilise, stablise and combine a multitude of elements (Murdoch & Ward 1997: 311). However, rather than simply reflecting or transmitting abstract rationalities, the practices, techniques and devices through which government is conducted are active in 'co-ordinating people and things in new configurations' (Timmermans & Epstein 2010: 84) and in turn constituting 'the domains which are to be governed' (Murdoch 2000: 513). This concern with how discontinuous elements are held together has drawn analysts to consider the forms of commensuration that are required alongside the deployment of forms of calculation, examining the means through which alignment is not only sought but achieved in practice and sustained. Here too there has been a focus on those devices that enable abstraction and enable 'making things the same' (MacKenzie 2009). Predominantly concerned with the making of markets, such work has drawn attention to the practices of measurement and accountancy in producing forms of value that can be combined and exchanged, working in specific marketplaces and across time and space (Blok 2014; Lansing 2012; Lovell & MacKenzie 2011; Lovell 2015).

This chapter seeks to examine how calculation and commensuration have come to be deployed and enacted through the governing of climate change in the United Kingdom. Analysts have drawn attention to the ways in which such techniques have been critical to the working of carbon markets (Callon 2009; Lovell 2014; Lovell & MacKenzie 2011;

MacKenzie 2009), the government of forests (Lansing 2012) and development of individualised responsibilities for responding to climate change (Paterson & Stripple 2010). The first part of the chapter turns to matters of calculation and examines the ways in which techniques, practices and devices were mobilised in order to establish the feasibility and value of interventions designed to work on climate change. Such forms of calculation were not only quantitative but also pertained to the qualities that could be realised through working on climate change, seeking not only to operate and mobilise governing at a distance but also through the proximities and contingences of particular sociomaterial conjunctures. The second part of the chapter examines the ways in which commensuration has been sought, achieved and contested as the work of governing climate change has come to ground in particular communities, businesses and places. Here, it finds that 'making things the same' has been achieved through techniques, practices and devices that enact and enable a 'logic of equivalence' (Li 2011) such that new entities and values mobilised through governmental programmes are entrained within existing configurations. These forms of governmental work come under pressure as the specificity of climate change responses needs to be retained, certified and exemplified and as equivalence is questioned and matters of incommensurability overflow the intervention.

Making Climate Change Count

A substantial body of work drawing on Foucault's concept of governmentality has focused on what Rose-Redwood (2012: 300) terms 'the politics of number, or the calculative rationalities of political practices' interrogating such things as 'the role of statistics, census-taking, and map-making as technologies of power that developed in conjunction with the rise of the modern territorial state' in order to examine how 'governmental power is exercised through the deployment of political technologies of calculation'. Early work in this field, such as that of Rose and Miller (1992: 185), drew on Latour's actor–network theory to examine the ways in which inscription devices served to make 'reality ... stable, mobile, comparable, combinable' and were central to the formation of "*centres of calculation*" (emphasis in the original: see also Murdoch 1997, 2004; Uitermark 2005; Walters 2012). Whilst at pains to point out that 'this is not a matter of the implementation of ideal schemes in the real, nor of the extension of control from the seat of power into the minutiae of existence' (Rose & Miller 1992: 183), many could be forgiven for reading accounts of the exercise of governmental techniques of calculation in this way (Allen 2003). In their detailed analysis of the developmentalities afforded through the UN Millenium Development Goals, for example, Ilcan and Phillips (2007) examine the ways in which a calculative toolkit involving statistics, audit, information profiling and so on serve to create particular ideas about who and what are responsible for development. Like others who deploy such approaches, they examine the ways in which techniques and practices are transmitted from the centre through and on to a range of different sites and subjects in order to achieve particular forms of development. Other research, however, suggests that any such apparent distance between centres of calculation and the sites through which government is practiced need to be questioned. Rather than being an automatic or readily transmitted entity, the provisional achievement of authority is dependent on the translation and continual incorporation of a multiplicity of agents into governmental programmes that in turn serves to produce the 'effect' of the state or other centres of calculation (Murdoch & Ward 1997). Rather than taking the 'programmers view', of seeing power from above and governmentality as a mode of government to simply be rolled out and orchestrated above or across a social body (Death 2014), attending to calculative practices of assemblage provides an alternative means through which to consider the work that such techniques and devices perform.

Performing Calculation and Carbon

Rather than being concerned only with the ways in which the state deploys such techniques, studies of governmentality have expanded to examine the multiple sites through which power is manifest. Together with the ongoing interest across science and technology studies with the artefacts and devices through which assemblages are constituted have led to a series of new engagements with how and on what basis calculation comes to matter. This work suggests that far from providing a set of rules or guides for the conduct of individuals, calculation in its many guises functions as one means through which governmental assemblages are articulated and given meaning (Friedberg 2013; Loconto 2014). In this sense, the techniques, practices and devices of calculation serve to refine and compose the field of intervention whilst also ensuring that governmental programmes come to both be calculable and calculated (Li 2007a: 6). Making government calculable requires identifying and composing the ways in which the goals of improvement can be identified and limited – that particular finalities are created. Making government calculated entails ensuring that the weight or importance of governing is recognised and its worth valued. As Lansing (2012: 207) argues, 'crucial to this view is the idea that acts of calculation are at once abstract representations and practices that are imbued with a materiality ... [such that they] become material interventions into how [for example] economic action unfolds' (Lansing 2012: 207).

Central to this work has been a concern with understanding the ways in which markets are realised and performed (Callon 2009). For example, in her analysis of the marketisation of the Bristol Bay salmon fishery, Hébert (2014: 22) shows how 'rationalization and niche marketing projects engage market devices such as salmon catch shares and fillet quality standards to realize widely circulating economic models', yet in practice such devices serve to 'reveal both the workings and the failures of performative reordering' such that whilst markets are reconfigured these processes are 'never quite as anticipated.' Another area of focus has been on the role of standards and certification. Instead of viewing standards simply as rules for regulating behaviour, this literature suggests that they need to be regarded as 'sociotechnical devices that are used to discipline and govern humans, things and the processes that bring the two together'(Loconto 2014: 3). Standards attain governing power through a panoply of infrastructures and artefacts, including forms of institutional practice including standard setting, accreditation, classification, evaluation on the one hand, and the design and deployment of particular entities, such as codebooks, recycling bins, electric vehicle charging points and so on, through which standards are maintained (Lovell 2014; Timmermans & Epstein 2010).

Market making and standard setting also preoccupy accounts of the calculative work of governing climate change (Lovell & Mackenzie 2011). Central to this body of scholarship has been analysis of how carbon comes to be accounted for and given value. For some scholars, indeed such is the importance of such techniques that 'contemporary climate governance hinges on the ability to account for stocks and flows of carbon' (Lövbrand & Stripple 2011: 187–8). Accounting for carbon comes in multiple forms, prominent amongst which Lövbrand & Stripple (2011: 188) argue are '"the carbon sink", "the carbon credit" and "the personal carbon budget"'. Nonetheless, it has been with forms of carbon credit and the markets that they make possible that much of the research in this domain has been concerned. Such work requires a host of mechanisms, and practices are learnt and deployed, for 'the political decision to create a carbon market such as the CDM [Clean Development Mechanism] or ETS [EU Emissions Trading Scheme] is not the same as constructing such a market. A new commodity – an emission allowance or emission credit – needs brought into being: defined legally and technically, allocated to market participants, made transferable and tradable, and so on' (MacKenzie 2009: 444). For MacKenzie (2009), at the heart of such market-making practices is the notion of 'sameness' such that tradeable allowances can be made 'fungible', that is, comparable and combinable, between different greenhouse gas emissions as well as over time and space. This, as he goes on to elaborate, requires the complex attribution of 'global warming potential' to different gases and the stabilisation of these inscriptions. Yet such market-making practices also rely on calculative practices and techniques that are far from stable. In her account of the work of the International Accounting Standards Board (IASB), a body central to the governing of the ETS through advising on the process through which emissions should be accounted for and disclosed, Lovell (2014) demonstrates the highly unstable and contingent nature of accounting for carbon, as in the absence of international standards a multiplicity of different devices and techniques emerges.

In multiple and contested ways, therefore, 'not only do practices of calculation constitute what the carbon commodity is . . . but they also bring into being ways in which the carbon commodity can circulate and become useful to others' (Lansing 2012: 207). In parallel to the making of carbon markets, various forms of calculation have also been central to what Paterson and Stripple (2010: 347) term the 'conduct of carbon conduct' for individuals, corporations and communities. As governments and other agents have sought to act on and through low carbon subjectivities in order to address growing carbon emissions, techniques that seek to account for the carbon emissions of everyday life and enable individuals, organisations and collectives to either address these directly or offset their emissions have also grown in popularity (Lovell et al. 2009). Carbon footprinting, in which calculations of the carbon imprint of (a proportion of) product lifecycle are attributed to specific goods and services 'has become perhaps the elemental form of practice as individuals self-govern in relation to climate change' (Paterson & Stripple 2010: 347). As with other standards, such footprints come to have force not only because they codify carbon in particular ways but because they serve to constitute what it is that should be considered as within the bounds of carbon government and that which lies beyond.

Calculation in its various guises is then central to the assemblage of governmental interventions and the pursuit of the will to improve climate change. Such practices, techniques and devices work to further draw disparate elements into conjunction and enable forms of alignment that can become taken for granted, fading into the background and providing the vehicles through which government is conducted. Yet at the same time, they can expose the deficiencies and limits of such technical orderings, bringing into question once again what it means to enact and respond to, in this case, climate change. In what follows, the means through which calculation has come to matter in accomplishing climate governance are explored through the case of the Berwick CoRE wind turbine. The case shifts the focus of carbon calculation from a concern with the making of markets and the practices of exchange towards other arenas through which carbon's possibilities and value come to be conceived and configured. The case points to the ways in which calculation is not necessarily tied up with abstraction and simplification, such that 'actors become temporarily cleaved from their previous webs of relationships and entanglements' (Lansing 2012: 209), but may rather depend on bringing such entanglements into the calculation of carbon's worth.

Installing Wind Power in Berwick-upon-Tweed

Since the late 1990s, the development of renewable energy technologies in the United Kingdom countryside has attracted both vociferous support and significant controversy. One means through which the United Kingdom government sought to accelerate the deployment of such technologies has been through the engagement of local communities. Through the 2000s, several schemes were advanced by which the United Kingdom government and a range of associated bodies made funding available for the development of community-renewable energy schemes of various shapes and sizes (Walker et al. 2007). Berwick CoRE, a social enterprise formed by the partnership of CoRE (Community Renewable Energy) and Berwick Community Development Trust, was formally established in 2007 to progress one such project: the Berwick Community wind turbine (see Chapter 2). Installing wind power in Berwick involved not only the sorts of translation, alignment and comprehending of climate change discussed in Chapter 4, but critically forms of calculation through which this particular governmental intervention came to be counted as both feasible and valuable.

In Berwick, initial interest in renewable energy was generated through a rural renewable energy conference organised in 2005 by the local strategic partnership. With standing room only in the village hall, the organisers realised the significant appetite but lack of capacity for communities to organise their own energy provision. Importantly, this capacity was not innate to the community, or indeed to the forms of energy they sought to produce, but emerged in relation to the sorts of calculation required to bring such interventions to fruition:

> . . . we came to the conclusion that . . . when it came to the community being given a feasibility study, it just stalled all together. And the reason it stalled was because the community hadn't got the time, or the expertise, to take it forward. So if it was just photovoltaics on the village hall, they might just get there. But if it was anything larger like digging up the playing field next, for a ground source, or putting in a wind turbine, forget it (Berwick Interview B 2011).

Calculating the feasibility of renewable energy projects is an arduous process. It requires multiple forms of assessment – forms of assessment that have not necessarily been developed with renewable energy, carbon or climate change in mind. Leaving it to the community as an inchoate and loosely articulated assemblage to undertake such forms of valuation was not sufficient. Rather, a specific vehicle had to be constituted that was capable of marshalling wind energy through the process of its calculation. The formation of such an entity was in itself a calculated act, requiring the alignment of community enterprises (CoRE and the Berwick Community Development Trust) with available resources:

> . . . the next thing that we did was snaffle some of the north east social enterprise partnerships 'Inspire' money. 'Inspire' was a European social fund project . . . we snaffled some of their funds, we spent nearly a year producing a business plan (Berwick Interview B 2011).

Creating a feasible wind energy intervention required the presence of a suitable vehicle – one which met the requirements of funders for sustainability and viability through the production of a particular sort of standardised and standardising device, a business plan. With a business plan in place, CoRE could be incorporated as a company limited by guarantee and the partnership between CoRE and the Trust could be established. This corporate and legal architecture in turn provided the basis for further forms of calculation to be undertaken. To begin with, the capacity of the turbine itself has to be determined. This is not a calculation that can be made on paper or in isolation, but requires the imbrication of the turbine with its situation. The initial assessment of the site for the location of the wind turbine was a far from technical judgement. A previous proposal for the development of a wind farm near the city had been rejected and in its place an extension of the business park proposed. Recognising the opportunity, those involved with CoRE advanced an application for one turbine to be included as part of the development of the land. When this was rejected, the option of putting a turbine in another nearby location – a field – was explored with the Berwick Freemen, a powerful economic group in the city, which granted CoRE funding to 'buy and install a net mast' through which the potential of the site could be measured and monitored. This fifty-metre wind-monitoring mast was then erected in 2008 to gather data about wind speeds. In 2009, external consultants verified that the site was not only appropriate for a turbine, but also that it was commercially viable:

> . . . you can see from the larger geography around that the wind funnels through places. We needed seven metres per second to make it commercially viable, and we've got nearly eight. So it's not a straight line, the more wind you've got, it's actually a logarithmic progression and the energy that you get in return. So instead of being commercially viable, and a little bit more, we're actually commercially viable and a whole heap more (Berwick Interview B 2011).

Technical calculations are of course not sufficient to bring forth governmental interventions. As the preceding quote reveals, it was the economic calculation implicit in how the wind blew that mattered. As the viability of the intervention became more assured, Berwick CoRE sought additional funding for the capital costs of installation through the government's Low Carbon Communities Challenge (LCCC) programme (see Chapters 2 and 3). All too mindful of the emphasis of the Challenge on the effectiveness of 'integrated community support packages', the Berwick wind turbine became reconfigured as part of an amalgam of low carbon interventions designed to support Low Carbon Berwick, including the installation of photovoltaics on community housing and measures to reduce energy demand and tackle fuel poverty. The bid to the Department of Energy and Climate Change (DECC) for £0.5M to support the capital costs of both installations split the request 50:50. Following the application stage, short-listed projects hosted an appraisal visit from DECC. The rational calculus applied by DECC was such that that it was unable to fund the part of the proposal that was associated with the community wind turbine as, at that point of time, the project did not have planning permission:

> They did an interview with us and said 'we're not really sure about the wind turbine because you haven't got planning yet, so we'll just give you a quarter of a million for the PV'. Which we haven't got planning for either (Berwick Interview B 2011).

Yet the qualities that the wind turbine stood for and embodied – the partnership and community engagement it had engendered, the sense that Berwick was a 'first mover' (to whom the scheme was particularly targeted) and the vitality of data concerning the viability of the wind scheme – were also drawn into the calculation that Berwick's low carbon project was both feasible and desirable, securing its participation in the Challenge scheme (Paterson & Stripple 2012). Participation in the Challenge scheme led to a new range of calculations beyond the technical and economic feasibility of specific installations and towards forms of assessment and evaluation. The concern with assessing the ways in which the presence of the scheme had (and had not) led to changes in attitudes, behaviour and the production of carbon across the community was regarded as intrinsically difficult to calculate. Yet, given the expectations of the scheme, and the forms of calculation it was seen to embody, the relative lack of engagement from DECC with the subsequent fate of its heir project finance came as a surprise:

> We've not delivered for solid technical reasons. ... But I don't get requests for progress reports either. I don't get anybody ringing up saying, 'Julian, how's it going? Have you spent the money, blah, blah, blah?'... Now, in some ways it's great to not have somebody on your back, and we are working hard on it, and we have a path that will lead to installation, but it's been pretty surprising how relaxed they've been (Berwick Interview A 2011).

In DECC's own evaluation, Berwick Core is listed as one of four projects that were not able to complete their projects 'to the LCCC timeline' (DECC 2012: 10).

With funding from DECC to pursue other low carbon activities, the question of how to create the capital for the wind project remained unresolved. Traditional forms of capital finance from high street banks were seen as challenging because of the number of different calculations required, from 'planning permission, a lease, a grid connection, [to] a power purchase agreement, all that sort of thing' (Berwick Interview B 2011), before the project would be considered feasible. Alternatives were, perhaps ironically, deemed incalculable. On the one hand, it was not legally possible to give away shares in a community-owned scheme, but rather they would need to be bought, potentially limiting the participation of those with low incomes. On the other hand, the limits on the size of such investments meant that in a small community such as Berwick the possibility of leveraging the required capital was limited. Such financial calculations were then intimately linked to the ways in which other forms of feasibility were assessed. Here, the planning system also proved an obstacle, not only because of some opposition to the scheme that was given voice through the planning process, but because the forms of assessment available were intended for a wind *farm* rather than a single turbine, resulting in the sense that the 'planning [system] doesn't understand what we're doing' (Berwick Interview B 2011). New forms of agreement between the farmer who had given permission for the turbine and landowners, the Berwick Freeman, also had to be negotiated.

In 2012, funding for the turbine was announced in the press. Verco-Baxendale, a climate change and energy consultancy, decided to invest in the turbine at a cost of approximately £1.45M with financial backing from the Co-operative Bank and Baxi Partnership. The resulting income will be split, with a 39 per cent share of the profits going to Verco and the other 61 per cent to the Trust and CoRE (which have since been amalgamated) (Berwick Advertiser 2012). In May 2014, the Trust announced that the Turbine had finally been installed:

> On the 14th January the turbine was finally completed, it was commissioned in April and officially launched by the Mayor of Berwick on the 7th May 2014. After seven years of work by the Trust and all its partners the turbine is operating and generating clean sustainable electricity and an income to support the Trust's work. . . . The turbine will generate over the next 25 years an income of more than

£2.3m at today's prices. The 50-metre high 500 kilowatt turbine has a blade span of 24 metres and will generate annually the equivalent of 450 homes consumption. Extensive testing over some 15 months established that the windy site overlooking the North Sea is ideally suited as a location for 'green' energy generation with above average wind flows. Sixty one per cent of the lifetime income generated from the turbine will go to Berwick CoRE – a company now wholly owned by the Trust – with the remainder shared by the bank and other partners. The Trust will then use 80 per cent of Berwick Core's income to support its work in Berwick and 20 per cent will be used to support a grant fund enabling community energy projects across the North East (Berwick Trust 2014).

Even as it comes to fruition, Berwick's wind turbine is full once again of calculation, of its capacity and desirability, of the different values it embodies, and the ways in which this value can be recognised. Not only are the techniques, practices and devices of calculation multiple, but they are interwoven. In Berwick, as elsewhere, it is the interweaving of calculation, its repetition and refraction that serves as a means through which the governing of climate change is accomplished. At the same time, far from seeking to simplify and purify the assemblage, accounting for the co-benefits of the wind turbine, and indeed the wider project of being Low Carbon Berwick, requires that climate change is continually related to a multitude of other 'specific finalities':

> Most people seem to simply turn off when they hear 'climate change' . . . it's a much more positive attitude if you say 'you can make money from this', and they go, 'oh, right'. . .. And it's much quicker, and it's easy, and quite a lot of the people come to us on a climate change, you know, 'we want to do something about it', or whatever. But if you're going to engage the larger society, then, 'we can do this; and we can make money' (Berwick Interview B 2011).

In Berwick, making money was not an end in itself, but rather a means through which community trust could be developed and new forms of energy resilience developed. As Marres (2011: 514–15) argues, 'devices which ideally enable the calculation of carbon emissions . . . allow for multiple, diverging co-articulations of economy, politics and innovation, enacting the politics of contestation in a material modality'. What we have to contend with here is not so much a matter of overflow, of the things that escape and cannot be contained through calculative orders, the issues with which research in this field has traditionally been concerned, but of *inflow* of the things that want to be counted as climate change and which in turn serve to enlarge and animate its governmental programme.

Making Things the Same?

As the case of Berwick CoRE and the wind turbine suggests, government does not work in isolation; rather, a whole host of devices, techniques and practices are articulated in order to produce and sustain the field of intervention and the operation of governmental programmes. Alongside the importance of calculation, those concerned with the social production of markets have drawn attention to the equally important practice of commensuration: 'the transformation of different qualities into a common metric' (Espeland & Stevens 1998: 314). For Espeland and Stevens (1998: 316), it is this particular quality of the practice of commensuration, of bringing things that might normally be articulated in different form, into a common metric that puts it apart as a technique and that in turn means that it 'is noticed most when it creates relations among things that seem fundamentally different'. For the most part, much of the work of commensuration passes unnoticed, and yet the 'enormous organisation and discipline' required to achieve commensuration are 'crucial to how we categorize and make sense of the world' (Espeland & Stevens 1998: 314). Climate change is no exception. Commensuration is critical to how the notion of climate change has come to be categorised and made sense of in a range of different settings. Whilst common metrics, notably of carbon footprinting and accounting, have been critical to these processes, making climate change fit has required other forms of commensuration. The rest of this chapter considers how and why it might be useful to extend the notion of commensuration beyond metrics and standards, before turning to Tesco, the United Kingdom supermarket chain, as a case through which to examine commensuration in practice.

Coming into Agreement

> *... it doesn't actually matter whether a board believes in climate change, 'cos climate change believes in them (London Interview 2007).*

As with other governmental techniques, commensuration not only shapes the sociomaterial order, but through its constitutive force serves to configure entities and their conduct. Nowhere in the climate change domain is such work more apparent than in the arena of carbon markets. Alongside needing to calculate carbon, such markets have depended upon forging common metrics through which it can be traded and exchanged. In this context, 'commensuration allows for the creation of a commodity that is completely removed from physical materiality. . . . The methodologies and standards of emissions markets are concerned exclusively with measuring the CO_2 content represented by each unit of production, and making the value of the credit exchangeable with other commodities' (Knox-Hayes 2013: 124). Making carbon and other greenhouse gases into common metrics is critical not only in order to establish value, but also because it reduces the costs of each transaction. As McKenzie (2009: 446–7) explains 'liquidity in such markets would be greatly reduced if the relevant "exchange rate" between gases had to be negotiated ad hoc for each transaction.' The techniques, practices and devices through which commensuration takes place – whether that be through the use of measurement, indicators, models or record keeping – are both intended to foster coordination between individuals as they seek to evaluate their conduct in relation to what is required and to black-box the (often contested) negotiations through which such devices have been designed (Fernandez 2014). Commensuration, like calculation, is seen to work by ensuring that 'everyday experience, practical reasoning, and empathetic identification become increasingly irrelevant bases for judgment as context is stripped away and relationships become more abstractly represented by numbers' (Espeland & Stevens 1998: 317). In short, the force of commensuration works by serving to simplify what it is that is assembled in order to safeguard the smooth working of governmental programmes.

The black-box, taken-for-granted nature of commensuration has led Lovell (2014) to argue that in order to understand the ways in which it serves to configure the work of government, more attention should be paid to arenas in which new metrics, standards and such are devised and come to be fitted within existing orders. She suggests that 'whilst much of the literature on standards concentrates on the reach and powerful day-to-day operation of existing large bodies of standards (health, weather etc.)', much can be learnt from 'cases of things that do not fit with these existing standards' (Lovell 2014: 267). In short, Lovell argues that more attention needs to be given to those incommensurable entities that have trouble being accounted for and in turn trouble existing forms of accounting (Lovell 2014: 267). Incommensurability is not incidental, but is also a product and a force that is actively produced. Misfit is created for 'some party must draw boundaries around the thing whose value is to be kept, or made, distinctive and then defend the boundaries from encroachment'

(Espeland & Stevens 1998: 328). Illuminating forms of incommensurability therefore always draws attention to the political nature of governmental practices, techniques and devices and the ways in which they are contested and come to be undone. Whilst those adopting an assemblage approach to the making of markets do emphasise their fragile nature, such studies often focus on 'successful performances', leaving 'unanswered questions about why certain economic models seem to smoothly reconfigure the world in their image, whilst others gain little traction' (Hébert 2014: 21). Examining the failure of commensuration is important, for 'it may in fact indicate a set of more profound difficulties in the wider project and ambitions of government' (Lovell 2014: 279).

Attending to incommensurability also widens the scope of how we might conceive of what is commensurate beyond the calculative practices of metrics and the seeming intention to make things the same (McKenzie 2009). In addition to 'having the same measure', to be commensurate can also entail 'corresponding ... belonging to the same sphere or realm of things' and in its active form is defined as 'to agree or "square" with; to make to correspond in nature' (*Oxford English Dictionary* 2015). The literature on environmental philosophy speaks to this broader scope of (in)commensurability, attending to those things that can and cannot be made to agree or correspond – trees and money, time saved and landscapes destroyed. In its wider sense, then, commensuration is about the extent to which things do or can be made to *agree*, such that the techniques, practices and devices we might be concerned with move beyond those that seek only common metrics – critical as they are – and also include other means through which things are made to 'fit' with one another. In her analysis of the environmental conflicts emerging in and around a Chilean mining project, Fabiana Li (2011) argues for an expansion of the notion of commensuration to what she terms '*a logic of equivalence*'. As she argues, rather than being related through a singular or calculated metric, 'compensation agreements, environmental management and other CSR [corporate social responsibility] initiatives rest on ... a *logic of equivalence* that makes the possible consequences of a mining project commensurate with the mining companies mitigation plans.' In this sense, '*equivalence* draws on but seeks to broaden this conception of commensuration by examining other ways in which disparate entities are bought into relation' (Li 2011: 62, emphasis in the original). Incommensurability emerges as governmental techniques, practices and devices 'move across varied terrain' and 'the properties that are brought together in their creation do not always mesh felicitously with the arrangements [and entities] they encounter' (Hébert 2014: 23). In the Chilean mining case, attempts at enacting and sustaining the logic of equivalence came undone as the project of governing the environmental impacts of the mine encountered glaciers, which:

> came to represent an incommensurable value, something that could not be negotiated, and thus made the project inherently unviable. By changing the scale of the problem (making a local one into one of planetary dimensions), glaciers bought into view elements that disrupted the logic of equivalence, as well as new forms of ethical reasoning that extend beyond the usual concerns of CSR (Li 2011: 63).

Logics of equivalence are a form of commensuration sustained by multiform tactics and techniques that seek to fit and hold highly discontinuous elements together in order to achieve their correspondence and agreement, in turn serving as a means through which authorisation as consensus is accomplished (see Chapter 3). This offers an important means of expanding our understanding of commensuration beyond the metric, as well as the sites, of market formation through which it has been explored. This expanded sense of what it means for climate change to be commensurate provides fertile ground for examining how the will to improve climate change is expansive and should appear in the most unlikely of places. It also points to the various forms of work required in order that climate change comes to agree with existing sociotechnical configurations and political spheres. By the same token, incommensurability emerges not only in moments of high politics and open conflict, but through the manifold ways that acting on climate change fails to agree with established assemblages, whether these concern everyday mobility, the ways in which food shopping is conducted or energy generation. Examining the case of Tesco, one of the United Kingdom's largest supermarkets, the rest of this section examines the processes through which climate change came 'into agreement' with and through the corporation and the challenges that were encountered.

Every Little Bit Helps: Bringing Zero Carbon into Tesco

Tesco's engagement with climate change began in the mid-2000s as then–Chief Executive Sir Terry Leahy made a series of speeches pledging that

Tesco would not only see to reduce the carbon content of its business and customer practices, but intended to become a zero carbon business by 2050 (see Chapter 4). Addressing climate change was framed as a matter of reducing risk and taking responsibility, of developing new markets and of realising efficiencies:

> Reducing emissions is the right thing to do for a responsible business seeking sustainable profits. It conserves energy, saves money, helps deliver energy security and better resource efficiency (Tesco 2012a).

Positioning itself as one of a leading group of corporate actors in the United Kingdom responding to the Stern Review and aligned both with the nongovernmental organisations orchestrating a response, including the Climate Group and WWF, as well as with the United Kingdom's international position, Tesco sought to govern both itself and its constituencies through the setting of targets and a host of techniques, practices and devices designed to calculate and commensurate climate change within the business and its wider sphere. Carbon metrics were essential to this endeavour. Measuring the carbon footprint of the business, from head office to distribution, individual stores to particular products, became central to the work of bringing climate change into agreement with Tesco:

> As an organisation we are driven by data, we are driven by measureability and we are driven by targets. Focusing on carbon in the way that we've done allows us to do that (Tesco Interview A 2010).

> What carbon has allowed us to do is to do the stuff that we're really good [at] by giving everybody a very clear measurement ... one of the reasons why it's easy ... is because you can make a really clear financial case. So, you know, saving energy is easy because the store manager spends less money on energy ... it's not easy to do it, but making the case is easy (Tesco Interview A 2010).

Carbon became a unit through which comparisons could be drawn across the business and beyond; it formed a kind of currency through which the shared value of responses to climate change that exceeded the capacities of any individual firm could be articulated and agreed upon:

> ... there are a lot of fora in industry that look at how we can all benefit ... as businesses, but also to this concept of shared value, and carbon is a currency that you can agree on because it's across so many different issues. So you can use carbon as a proxy for deforestation, for example ... there are other biodiversity issues where you can't use it. But it provides a good common language. It really

does . . . people have got very used to the narrative of carbon. And so it is quite useful as a starting point to use that as a comparator for other sustainability issues (Tesco Interview A 2010).

Yet calculating and making carbon commensurable as a metric – a unit of value, exchange and comparison between different parts of Tesco's operation and through its supply chain – also required other forms of commensuration that enabled climate change to fit with existing business operations. In seeking to mobilise the target of reducing emissions from the distribution part of the business by 50 per cent, for example, Tesco needed to find a unit to which carbon could be attributed that was both standardised and made sense within the corporation:

And the way in which we decided to track that is by case delivered. Because, again, you can't account for change in volumes but what you can say is, the emissions associated with any case of goods, and since the shape and size of a case doesn't change (Tesco Interview A 2010).

Beyond the measurement and monitoring of carbon, a second form of commensuration involved practices of design. A key target for Tesco was that its new stores built between 2006 and 2020 would emit, on average, half as much carbon as that of a standard store designed in 2006. The building of low carbon stores provided a space within which Tesco could both demonstrate its intention to meet low carbon targets and experiment with the different means through which this could be met. In 2006, a new store at Diss, Norfolk, achieved a 50 per cent reduction in carbon emissions, and in 2007 a store with 70 per cent less emissions was opened in Cheetham Hill, Manchester, followed in 2009 by the first zero carbon store in the world at Ramsey, Cambridgeshire. Each of these stores incorporated features aimed at reducing carbon footprints, for example by building with wooden frames instead of steel; using natural light; employing alternative heating, cooling and ventilation systems that recycled air within stores; applying technical and behavioural measures to reduce energy and water consumption; and minimising energy use and leakage of greenhouse gas emissions through on-site energy generation and different types of refrigeration technology (Tesco 2010). From the outset, the technology within the stores was designed to be as unobtrusive as possible:

For customers, our new store at Cheetham Hill will be just as friendly and welcoming as any other Tesco store. However, while it might seem the same, it is actually very different (Tesco 2009a).

For the designers, the intention was not to draw attention to carbon – to make it stand out – but rather to make it fit within the existing Tesco store concept and experience, such that:

> what customers know is that it's a really pleasant shopping environment, and they really enjoy being there and they think they've been given a really smart shop. And that drives trade. So the intention is very much known and [. . .] is very much to mainstream green, and not to lose technological advance' (Tesco Interview A 2010).

At the same time, the low carbon features of the store are literally on display. Large glass frontage, wooden frames and low carbon technologies frame the entrance to the store, whilst inside the internal pipework of the HVAC system is laid bare, the usual chill of the fridge isles are tempered by the doors and signs proclaim its low carbon credentials. Carbon is clearly present, yet through being designed *into* Tesco, its presence is barely felt by those who work there or by consumers:

> . . . we don't really talk about it, it just feels like a normal Tesco really, apart from the doors on the chillers and the lighting and everything (Tesco Interview B 2011).

> I think I did notice the difference in the building that it was slightly different, I suppose I just thought it looked modern, oh, that's a nice modern Tesco's (Tesco Interview C 2011).

Bringing low carbon into Tesco's design meant not only experimenting with novel technologies and infrastructures but also attending to the local situation of each store. Whilst both Cheetham Hill and Ramsey have on-site Combined Heat and Power (CHP) units, at the former it runs on what is termed 'naturally produced fuel from sustainable sources', including the waste fat from the store's own chicken oven, whilst at Ramsey, the CHP plant runs on biofuel from fish oil and rapeseed oil, each reflecting the nature of the local biofuel economies (Tesco 2009a, 2009b). At Cheetham Hill, a wind turbine was installed, which required specific local knowledge to overcome the technical assessment that it would not be productive:

> Northwesterly winds are always blowing this way . . . It's not only knowing the place, it's also because of the technology for the first time here also, so we tried testing it and it actually worked. It actually worked. They had told us it wouldn't work, not one little bit, and it actually worked for the store (Tesco Interview D 2011).

Making carbon commensurate with Tesco's stores required then the interweaving of practices of design and local knowledge, both serving to signal their low carbon credentials and meld these entities into the configuration of a supermarket. This process was made explicit in the development of what Tesco terms its 'blueprint', the template from which all of its stores are developed, as a model for low carbon stores:

> Planning permitting, our new blueprint will be the model for every new store we build in the UK from now on. While we may not be able to use all these measures in every new store location, the blueprint ensures that all the new stores we build will have a carbon footprint at least 50 per cent smaller than our 2006 stores. We will continue to work on our new blueprint and hope to build stores with a zero carbon footprint. The new blueprint has been created to be flexible across all types of our stores, and elements can – and will – be incorporated into existing stores too (Tesco 2009a).

Explicit in this design device is that elements derived from the low carbon flagship stores can be incorporated into both new and existing buildings. This is unlike the case of domestic buildings, where analysts have shown how innovative housing designs have struggled to imprint on mainstream housing development (Lovell 2004). Critically, this is not a one-way flow in which novel technologies and practices are inserted into both new developments and existing stores. Rather, as technologies are put to (and pass) the test of delivering low carbon in a manner that fits with the Tesco's design, they are incorporated into the blueprint such that 'then each year the standard blueprint is, itself, lower carbon' (Tesco Interview A 2010). At the same time, through the retrofit of the existing estate, 'we may be coming up against solutions that we want to try as part of ... any new format' (Tesco Interview A 2010). So in turn being low carbon is made to come into agreement with the existing configurations:

> ... we try out everything, but we know what has been working in the zero carbon store, and then that goes back into our format plan. So, yeah, we've got specifications for all of the different site stores, and so the stuff that works go back into the standard blueprints. ... But equally, as we're retro fitting our existing [stores, we learn]. ... So it's not only one way. We use each other (Tesco Interview A 2010).

Alongside the commensuration afforded through the deployment of metrics and the techniques of design, a further set of what could be termed 'technologies of agency' (Dean 1999) were also critical to bringing carbon into Tesco. Such techniques seek to work through constituting subjects capable of governing their own conduct in relation to the programme of governmental intervention. Here, Tesco was concerned to govern carbon through the conduct of both its employees, in order to realise its own ambitious goals for delivering carbon reductions, and those of its customers, to meet its broader objectives of improving the climate change condition. At a store level, Tesco has implemented a programme whereby individual stores have energy champions, trained to make 'sure we are not wasting energy, and encourage their colleagues to do the same' (Tesco 2009a). Alongside the energy champions, each store has a display board monitoring and reporting progress against targets. Moreover, energy efficiency has been incorporated into the Tesco Wheel management system, which sets out key performance indicators across different areas of the business. Employees are directed towards these general business goals and given training about the specific areas of the stores where they work. Many employees in both Cheetham Hill and Ramsey felt that they had low carbon 'drummed into them' through these mechanisms, though for others their roles meant they felt distant from the often invisible technologies, with their aura of technical wizardry, and not necessarily part of achieving the store's low carbon credentials:

> I know a bit about it, only through my management training, but I wouldn't feel comfortable walking up to it and saying, this is that, it does that (Tesco Interview E 2011).

> 'I've noticed but I've never really asked, it's not a big thing to me (Tesco Interview F 2011).

Beyond the operation of the stores and the roles of staff within them, perhaps most visible of all of Tesco's efforts to bring carbon into its business has been the adoption of carbon footprint analysis for its products. Between 2008 and 2012, Tesco completed carbon footprint analysis of more than 1,100 of their own-brand products and used this information to carbon label more than 500 products, 'focusing on many of the products our customers buy every week, such as bread, milk, fruit and vegetables, together with some items that are known to be energy intensive, such as light bulbs' (Tesco 2012b). These products were chosen specifically in order to make sure that they were '*in as many baskets as possible.* That way it works' (Tesco Interview A 2010, emphasis added). By targeting the familiar and mundane products of the weekly shop, Tesco sought explicitly to bring carbon into the everyday practice of its customers. Carbon footprints, whilst not intended to provide expert knowledge, would enable customers to understand where they could, and could not, make choices that would shape carbon emissions. At the same time, footprinting also enabled Tesco to learn about the scope of its own carbon responsibilities and how these were distributed upstream, through supply chains, or outwards, through consumer practices. For example, their work with milk revealed that 'something like 70-odd per cent of the emissions are associated with the agricultural phase – i.e., the cows ... the methane. So that tells us if we want to try and reduce the emissions, then we have to do something about the agricultural phase' (Tesco Interview A 2010). Providing customers with carbon labels was seen as a means through which they could be empowered to take action through their own choices, an approach that fit squarely with the notion of consumer choice championed by the supermarket chain:

> So there's kind of the sense that small actions can make a difference ... *our whole approach to life* of 'Every Little Helps' is beautifully suited to it' (Tesco Interview A 2010, emphasis added).

Bringing low carbon into Tesco relied on the interweaving of metrics, design and technologies of agency that served to create equivalence between reducing carbon and developing Tesco, improving climate change and the business of corporate responsibility. Yet not all forms of being low carbon could be made to fit, and such agreement was often either temporally or spatially limited. The presence (and absence) of fridge doors is a case in point. Tesco, as is the case for most supermarkets in the United Kingdom, has traditionally used refrigeration units without doors. The low carbon stores in Cheetham Hill and Ramsey trialled new units with doors included, an adjustment that had to be accounted for not only in carbon terms but also through the management and working practices of the business:

> The company predicts how long everyone should take for a certain job. In this store, we have a higher number per cage [of products to be stacked on the shelves] because of the doors and that accounts then in our budgets for wages. We might have a slightly higher wage budget for the exact same store without doors. And they account for that (Tesco Interview E 2011).

Yet what could not be accounted for in such direct terms were the reactions of the employees to using such fridges. Whilst several found that it was another routine to get used to, and were impressed with the innovation and energy savings involved, others found the change to their working practices too much to adjust to:

> Awful. If you get a box you can't open the door and you have to put it down and the doors are slamming and customers are opening them and slamming them . . .. The worst really is they just won't stay open' (Tesco Interview G 2011).

The incommensurability of 'closed-door' fridges is felt not in moments of high politics or protest, but in the ways in which employees resist and contest the manner in which their working conduct is governed in a low carbon store and seek ways to 'work around' such new technologies. Unlike other aspects of low carbon stores that have been 'blueprinted', 'close-door' fridges remain off-plan, their future uncertain as they seem to clash with the working practices of employees and what it is that Tesco expects its customers to expect (though have not directly interrogated).

The decision in January 2012 not to pursue carbon labelling was a form of incommensurability that received considerably more fanfare than a few slamming fridge doors. This decision was described in the press variously as a *U*-turn or climb down (*Guardian* 2012). The complexities of developing beguilingly simple labels to communicate the carbon content of everyday products were such that they could not be made to fit within the business and practice of Tesco. On one hand, this was a form of accounting that was time consuming and expensive. On the other hand, it was a form that had also failed to be widely adopted across other retailers, such that Tesco felt that it was investing too much in relation to its competitors (*Grocer* 2012). As Friedberg (2013: 185) argues, such projects are not only technical but political, for 'participation in corporate and government-sponsored footprinting initiatives reflects more than an interest in measuring-to-manage. They also seek to manage the measures. In other words, they want a say in how the footprint of food products is defined, calculated and communicated. More broadly, they want to assure that the still-malleable metrics of food sustainability come to serve their own business models.' Lacking the potential to 'manage the measures' as other retailers withdrew from the field of intervention, footprinting became less viable as a means through which carbon could come into Tesco in a manner that would not be disruptive to its business. As carbon footprinting has disappeared from view, so to have intentions to address the carbon emissions of Tesco's customers, with websites and corporate reports seemingly silent on this original intention. Seeking to hold radically discontinuous elements together, devices such as carbon footprints or design blueprints are far from assured or predetermined, but rather unfold as the work of government is conducted.

Conclusions

> *Global warming in its current state is an issue that is unqualifiable, not in theory but in practice, for no framing is able to embrace it in its entirety. As the roots of the word indicate, an issue always finds an exit enabling it to overflow. It is protean, constantly changing as it spreads, irrespective of the frame into which we try to fit and enclose it (Callon 2009: 542).*

Whilst analysis of how climate change is governed has tended to focus on the issuing of decrees or the intricacies of discourse, this chapter has focused on work of a different kind – the forms of articulation required to accomplish the governing of climate change. Where such analysis has been developed in the literature to date, the focus has tended to be on market devices and the techniques of calculation and commensuration through which they come to configure the objects of government and the means through which they should be pursued even if, as the preceding quote from Callon (2009) suggests, such matters are difficult to pin down. Rather than being a fixed domain to be governed, articulating climate governance in this way is an immanent affair. Further, far from being a matter of 'rolling out' a suite of techniques derived from one or other political rationality (e.g., neoliberalism) towards particular ends, the intention of improvement is neither fully known or wholly fixed, and as processes of articulation unfold, new conjunctions emerge and the remit of governing climate change shifts. Importantly, the analysis here suggests that techniques and practices of calculation and commensuration are ongoing and take multiple forms. For example, the calculation of the wind turbine in Berwick involved assessments of the suitability of sites, of the generation of power, of the financial viability of the operation, of the return on government investment and of the possibilities of using different forms of energy in the community. It is through the articulation of these different forms of calculation that the governing of climate change as a matter of renewable energy in Berwick took shape and gained traction.

Understanding the importance of such devices and techniques in accomplishing climate governance requires that we go beyond the assessment of individual tools and consider how they are articulated.

Moreover, the analysis of how climate governance is being accomplished suggests that the working of such devices and techniques takes place both through technical registers – of metrics, assessments, indicators and so on – but also through being able to express and articulate other forms of value. Rather than working simply to abstract, techniques of calculation and commensuration are also used to embed, connect and draw into the realm of climate governance entities and arenas that might seem tangential or distant. Climate governance is not only a matter of *overflow*, of that which it cannot contain, but also of *inflow*, of that which is bought into articulation as a matter of climate concern. In terms of calculation, alongside new techniques of carbon accounting, long-standing assessments of economic viability, planning gain, environmental impact and so on are pulled into new alignments. Forms of commensuration work through logics of equivalence to bring existing business practices, community values and government endeavour into the climate change fold. For Foucault, government involved not so much 'establishing limits and frontiers, or fixing locations, as, above all and essentially, making possible, guaranteeing, and ensuring *circulations*: the circulation of people, merchandise, and air etc.' (Foucault 2009: 29, emphasis added). The circulatory, expansionary, nature of the will to improve serves to encompass an ever-wider array of things into the matter of climate change, and in turn to carry climate change concerns into a host of different milieu. Government is not only a problematising activity, but one which seeks to create new possibilities (see also Chapter 4).

The circulatory, expansive and unfolding forms of articulation through which climate change comes to count and be counted in turn mean that the issue becomes multivalent in its comprehension (see Chapter 4). Such devices and techniques 'enable the organization of spaces of multi-valent action, in which a routine act . . . is at once a technical, economic and ethical act' (Marres 2011: 520). Individual actions, technologies, entities and arenas are therefore neither purely economic or political, neoliberal or radical, but rather hold multiple equivalences, articulated and translated in relation to particular situations and the finalities towards which the intention to govern climate change is directed. The will to improve climate change and to govern, authorise and comprehend climate change becomes articulated in multiple ways, each of which has its own political possibilities. At the same time, the ongoing work of governing climate change in practice is beset with contradiction, contestation and incommensurability, as matters that stubbornly refuse to be concerned with climate change – from the design of fridge doors to the practices of supply chains – interrupt and resist such attempts to draw them into alignment. It is to the constitution of climate's publics, and their politics, that Chapter 6 turns.

6

Assembling Climate Publics

Introduction

Governing is a public matter. From diverse starting points, and with very different visions in mind, scholars point to the ways in which governing 'at whatever level of social organization it may take place, refers to conducting the public's business – to the constellation of authoritative rules, institutions, and practices by means of which any collectively manages its affairs' (Ruggie 2004: 504). For some, the public nature of governing comes from the actors and institutions involved – governing requires the kind of recognition and authority wielded only by state actors who, in turn, should be regarded as public authorities. For others, the public nature of governing pertains to the kinds of problems with which it is concerned. Here of course climate change stands out as a paradigmatic example of a 'public' problem (Paterson 2013). The framing of climate change as a public goods problem – where the causes are distributed and the benefits of action shared – is ubiquitous across the research and policy community. As Paterson (2013: 151) argues, 'from this framing, an enormous range of literature outlining the logic of climate change as a collective action problem has emerged, with the standard range of problems emerging – free riders and how to get round them, enforcement and compliance, and so on'. Taken together, these two dominant understandings of the ways in which climate change should be seen as public have been pervasive, leading to the development of international agreements and a strong focus on the role of the state in reducing GHG emissions and ameliorating risk.

Yet at the same time, the governing of climate change has come to trouble these accepted notions of the public and how it is constituted. As the multilateralism that has accompanied the notion of climate change as a public goods problem comes under pressure, a growing range of arenas and arrangements have emerged through which the governing of climate change is being conducted (Bulkeley et al. 2014; Hoffmann 2011; Paterson 2013). Climate governance exceeds the international regime to include 'both the co-ordination of states and the activities of a vast array of rule systems that exercise authority in the pursuit of goals that function outside normal national jurisdictions' (Rosenau 2000: 172). Moving beyond the state, research has revealed the manifold forms of private authority, public–private partnerships and alternative public authorities engaged in these forms of governing climate change (Newell et al. 2012). Rather than being attributable to the actors involved or to its commons qualities, here the public is envisaged as a domain: an arena where 'expectations regarding legitimate social purposes, including the respective roles of different social sectors and actors, are articulated, contested, and take shape as social facts' (Ruggie 2004: 504). The publicness of such forms of governing climate change arises from the ways in which they serve to generate authority and legitimacy.

Paterson (2013) argues that such an account of the public as a domain or sphere requires a shift in our understanding of how and by whom climate is governed. He suggests that the emergence of climate governance experiments that are hybrid and operate across multiple scales (Hoffmann 2011) has been accompanied by their contestation as spaces for legitimate forms of governance because of the ostensibly *private* character given by the participation of certain actors. Understanding such forms of governing as public requires, he argues, that we not only open up to scrutiny what is meant by the public – as actor, issue or domain – but also enquire as to how it is practised. Here Paterson's (2013) analysis insists on a move beyond the conventional tropes of legitimacy that are regarded as making governing accountable in representational terms and instead turns to classic accounts of the public and the work of Hannah Arendt to argue that what constitutes the climate (re)public is made through practices of deliberation and intersubjective learning. In reconstituting the climate public as something that is produced through its practice, Paterson's analysis is located with those who resist any 'reductive dichotomy of public/private' and seek to account for how publics (and in turn what is held to be private) are a product of ongoing political processes (McGuirk & Dowling 2009: 177; see also Bulkeley & Schroeder 2012). Climate publics emerge not in the classical conditions of the agora, but in the complex, fragmented and intermeshed sites and spaces of governing climate change through the practice of making climate change public: of creating the deliberative spaces through which the governing of climate change unfolds in more or less consensual or agonistic ways (Paterson 2013).

This chapter begins from this recognition – that it is not the public that creates climate governance, but that governing makes climate public. Rather than being given, climate's public has to be made, fashioned through the ways in which governing is accomplished. Drawing attention to the multiple and various forms of climate re-public emerging at different sites and scales shifts our attention away from the conventional analysis of climate's public as the international community of nation-states and beyond questions of representation (Paterson 2013). Yet such domains appear, as their conceptualisation in classical notions of political deliberation suggests, largely outside of the dominion of private life and primarily orchestrated by forms of discursive and intersubjective exchange (Marres & Lezaun 2011; Rubio & Fogue 2013). Alternative perspectives drawing on Dewey's conception of publics as being formed around matters of concern suggest a need to move beyond the making of climate *re-public* to consider how climate serves to constitute *res-public*, for as Latour (2005: 16) articulates: 'procedures to authorize and legitimize are important, but it is only half of what is needed to assemble. The other half lies in the *matters* that matter, in the *res* [things] that create a *public* around it. They need to be represented, authorised, legitimated and bought to bear inside the relevant assembly.' Taking his inspiration from the linguistic roots of the word *thing* as a space of public assembly in Norse cultures (Telier 2012: 1), Latour argues for the inseparability of the making of publics and the things through which they are constituted. The first part of this chapter explores the implications of understanding the making of climate publics as taking place through the registers of everyday practice, whether this be in the home, the office, the street or elsewhere, and as configured through the ways in which matters are brought both to be of concern and into order (Pellizzoni 2012). Practices of government entail traversing the boundaries of ostensibly public or political realms and those orchestrated around the 'material reproduction of everyday life' (Marres & Lezaun 2011: 491), as conducts and subjectivities are enrolled into the workings of governmental programmes and in turn reconfigure how governing is conducted. Such processes are not confined to the workings of discourse or rationalities, but are configured through the ways in which different sociomaterial entities are related as matters of climate concern as 'objects, devices, settings and materials, not just subjects, acquire explicit political capacities, capacities that are themselves the object of public struggle and contestation, and serve to enact distinctive ideals of citizenship and participation' (Marres & Lezaun 2011: 490). The second part of the chapter examines the ways in which making climate public takes place in practice through examining the cases of Hexham Hydro and Brixton Low Carbon Zone. Whilst writing on the public as orchestrated by 'matters of concern' has tended to focus on sites and moments of controversy, this analysis suggests that forms of consensus are equally animated by dynamics of concern – of regard, of investment and of care.

Making Things Public

The question of how to make climate change public has been tackled in many ways. Most common have been those efforts focused on educating the public as to the nature and effects of the changing climate and how they might seek to moderate their behaviour in response (Shove 2010). Largely in parallel, research and policy communities have also sought means through which to make decision making in the international and transnational climate governance arenas more public through forms of institutional design that foster legitimacy, transparency, learning and so forth. Elsewhere, celebrities have sought to campaign for climate change, protest marches have taken place and a host of deliberative forums and policy processes have been designed in order to draw the public into climate change (Boykoff & Goodman 2009; Hobson & Niemeyer 2011; Saunders 2012). For the most part, climate change comes to be public in these accounts as and when it makes an appearance in what are regarded as the spaces of the public sphere – the policy process, the street, the media. Given that governing, 'at whatever level of social organisation' it takes place, is concerned with 'conducting the public's business' (Ruggie 2004: 504), the ways in which we understand the climate public has been central to both the conception and articulation of its governance. Yet there are at least two ways in which attributing climate change as public in this manner limits our understanding. First, the assumption that the space of climate's publics can be neatly cleaved from the ostensibly private sites through which the practices that produce GHG emissions and various forms of climate vulnerability serves to delimit any analysis of how climate is governed in rather unhelpful ways. As the previous chapters have demonstrated, such boundaries are far from fixed, and programmes intended to improve climate change readily transcend taken-for-granted distinctions between public and private realms. Second, such a conception of the public has tended to focus on its discursive and intersubjective constitution whilst neglecting the 'constitutive role that different materials play in constraining, allowing and informing the ways in which publics are drawn together and separated' (Rubio & Fogue 2013: 1041). Emerging from science and technology studies, as outlined previously, alternative perspectives argue that 'publics ... cannot be disassociated from the specific material and natural environments, technologies and infrastructures in and through which they are constituted and mobilized' (Rubio & Fogue 2013: 1041). Attending to the ways in which climate publics are made requires, therefore, that we look not only at their constitution in terms of the actors, discourses and practices involved but to the material means through which they are assembled and come to have presence and meaning.

Making the Private into Publics

Across the literature on climate governance, it is increasingly apparent that 'citizenship is no longer primarily realized in a relation with the state, or in a single "public sphere", but in a variety of private, corporate and quasi-public practices from working to shopping' (Pellizzoni 2012: 225). As Paterson (2013) argues, traditional assumptions about the whereabouts of the public that locate it in a discrete sphere fail to account for the manifold ways in which forms of intersubjectivity that constitute public practice now take place in relation to climate change. At the forefront of such analyses have been forms of governance arrangement that appear, at least at firsthand, to be private in character – alliances between corporations or partnerships between nongovernmental organisations and the corporate sector (Bulkeley & Newell 2015). Once regarded as operating outside what might be conceived as the legitimate sphere of climate governance, such entities and arrangements are increasingly seen as part of the complex through which governing takes place (Abbott 2012) and attention given to the ways in which they gain authority and seek legitimacy (Paterson 2013). Much of the literature has tended to examine such arenas as distinct spheres of private governance and to interrogate whether or how they might be made to (appear to) be more legitimate (Bulkeley & Schroeder 2012). Yet as Sending and Neumann (2006: 658) argue, the critical point is that such forms of governing do not replace those conducted by public authorities, but rather trouble any a priori division between the public and the private, as seemingly non-state entities come to do the work of governing in the public interest.

Similar forms of boundary trouble can be found in the imagination of the climate public in accounts that tend to regard it as somehow discrete from the 'the extra-political spheres of domestic life, work, leisure or economic action' (Marres & Lezaun 2011: 492). Such perspectives neglect the ways in which the governing of climate change entails not only forms of intervention designed to draw individuals into an ostensibly public realm but also those that seek to intervene in the intentions and practices of individuals. Conceived in terms of a governmental will to improve, the governing of climate change takes shape through acting on the conduct of subjects – of entraining the actions of everyday life into the various programmes and interventions through which governing climate change takes shape through structuring 'the field of possible action of individuals' (Paterson & Stripple 2010: 346). Rather than being located in a designated space or defined by particular forms of legitimate political practice, climate's publics are assembled through the making and collecting of subjects willing and able to participate in the conduct of governing climate change 'through certain forms of knowledge (measurements and calculations of one's own carbon footprint), certain technologies (the turning of carbon emissions into tradable commodities), and a certain ethic (low-carbon lifestyle as desirable)' (Paterson & Stripple 2010: 347). As Paterson and Stripple (2010) go on to argue, this is not a form of individualisation devoid of politics but what they term a kind of 'My Space', a virtual yet connected public realm through which the politics of responsibility for action are played out, resisted and refashioned. In one example of this form of government, Hargreaves (2014: 127) analyses the ways in which the rollout of smart meters and real-time display units across homes in the United Kingdom, planned for completion in 2020, are intended to help individuals 'control consumption, save money and reduce emissions' an effort by the [Department for Energy and Climate Change] 'to introduce a particular expression of carbon governmentality ... beyond the state and right into the heart of citizens' lives'. The climate public more or less successfully assembled is one constituted in relation to particular rationalities of how the governing of climate change should be conducted as articulated through the workings of particular governmental techniques that in turn serve to place individualised subjects as responsible for attaining the public good.

Viewed in this manner, climate's publics can be seen in quite different terms – as constituted not in those spaces which a priori are designated as arenas for public assembly, but rather as assembled through the practice of governing itself. Foucault's notion of the constitution of the subject and the population has been critical in this regard. As Marres and Lezaun (2011: 494) suggest, 'it is impossible to exaggerate the impact of Foucault's work on how the very question of the physique of the public – the manner of its structuration by material, environmental or technical means – is posed today', for work on the 'constitution of publics is grounded in a Foucauldian sensibility towards the "micro-physical" workings of modern governmentalities'. For some Foucauldian scholars, it has been the tendency of programmes of advanced liberal government to constitute publics that enable this particular political ideology to be 'rolled out' that has caught attention. Acting at a distance, centres of governmental power have developed rationalities and techniques that are apparently able to bring the behaviour of individuals and communities into line with governmental intentions in a rather straightforward manner (Dowling 2010; McKee 2011).

Yet the very idea of government at a distance is highly problematic (Allen 2003; Blok 2014; Pellizzoni 2012). Adopting a programmers view (Death 2014), such accounts tend to 'promote an overly abstract view of governing in which politics is reduced to rationality . . . [and] contributes to a representation of power that is omnipresent and totalising: thereby precluding the possibility of meaningful individual freedom and human agency' (McKee 2011: 13). Moreover, such accounts of governmentality as the direct articulation of subjects through the application of particular rationalities stand in contrast to other interpretations of Foucault's original interests. In particular, his understanding of the ways in which the operation of power depended on the condition of freedom such that 'individual or collective subjects . . . are faced with a field of possibilities in which several kinds of conduct, several ways of reacting and modes of behavior are available' (Foucault 2000a: 342). Other neo-Foucauldians adopt such an interpretation such that the subject is 'not framed as the negative effect of power wielded by the state or other institutions but as a productive dimension' of the ways in which governing is accomplished through 'the arrangement and management of people and objects, producing the material conditions that make particular ways of being comprehensible and possible' (Gabriel 2014: 42). Far from being assured, governing as the orchestration of a will to improve through the assemblage and configuration of particular dispositions is uncertain, unfolding, open to contestation and alteration. Climate's publics are in this context best conceived as formed 'in the making' and held together in provisional ways creating (new) sites through which what it means to govern climate change is both enacted and disputed.

Matters that Concern

If climate's publics are not given in advance of the intention to govern but generated through the ways in which governing is conducted, the way in which climate comes to be understood as an issue that requires intervention is critical to the constitution of its public (see Chapters 4 and 5). Such an approach accords with the pragmatist John Dewey's account of the public as a group that 'together are influenced or impressed upon by a specific set of conditions. As a public, these individuals seek to address these conditions and their consequences. These issues can either be of the present or the future; they can be experienced or anticipated conditions and consequences' (Le Dantec & DiSalvo 2013: 243). Latour (2005: 14) takes up this notion of the public to argue that it is such 'matters of concern, the issues we care for', that serve as the means through which publics and their politics are assembled. Such issues are not immaterial but rather come to matter in the shape of so many objects, each of which 'gathers around it a different assembly of relevant parties. Each object triggers new occasions to passionately differ and dispute', in turn binding us all in ways that map out a public space profoundly different from what is usually recognised under the label of 'the political' (Latour 2005: 15). Here, the space of the political is situated in relation to the translation of particular conditions into 'matters of concern' for the public that is so assembled (Blok 2011: 457; Whatmore 2013: 38). Moreover, the assemblage of publics is not confined to actors and discourses, but is both orchestrated around particular forms of sociomateriality (the matters of concern) and constituted through a heterogeneous collection of social, technical and natural elements. Accomplishing climate governance from such a reading entails rendering particular conditions (present or future) into concerns and the practice of assembling a public capable of forming and sustaining shared attachments to such matters.

Drawing on Foucault's notion of government as concerned with a 'sort of complex of men and things' (2009: 96), accomplishing climate governance can be seen as entailing the reconfiguration and contestation of different kinds of sociomaterial order. Yet there is a sense in Foucault's writing and those who adopt its conventions that governing is a set of technical practices through which particular problems come to be understood and acted upon as a matter of course. For Marres and Lezaun (2011: 494), such approaches render the materiality of politics too covert, tacit and controlling, such that:

> . . . in the neo-Foucauldian idiom, political subjects and spaces are configured through the deployment of objects, devices, environments and infrastructures, but this is a process that typically occurs informally and unobtrusively. Material elements operate in a "sub-discursive" manner: their relevance to political processes remains unacknowledged or at least under-articulated in public discourse (Marres & Lezaun 2011: 494).

However, such a reading of the political possibilities of the government of 'men and things' is, as with much of Foucault's work, only one lens through which it could be interpreted (Walters 2012). Elsewhere, Foucault's writing points to the ways in which the conduct of conduct at the heart of the governing of sociomaterial assemblages affords agonism between the will to govern and the condition of freedom that is at once its condition and its limit (Foucault 2000a: 342). As Li (2007a: 10–2) suggests, Foucault insists on the interrelation between securing particular forms of government and opening them to critique, such that this is a relation of 'permanent provocation' (Foucault 2000a: 342) in which 'between a relationship of power and a relationship of struggle there is a reciprocal appeal, a perpetual linking and a perpetual reversal' (Foucault 2000a: 347). This form of provocation may not be overt in the terms demanded by those who theorise the political as a site of protest (Swyngedouw 2010). Rather than being due to a lack of struggle or the under-articulation of the ways in which the practice of governing is political, the character of provocation may instead be related to the different modes of power at work and the ways in which, for example, authorisation, seduction or manipulation are deployed and challenged (Allen 2003; Li 2007a).

Yet for Li (2007a: 12), as for others, it remains helpful to distinguish the 'practice of government', the rendering technical of particular problems and their mobilisation through different forms of conduct to multiple diverse ends, and the 'practice of politics' where both the means and the ends of government are held to critique. An alternative, explored in this chapter, is to engage the notion of the matters of concern in order to analyse how, why and by whom the 'practice of government' is at once the 'practice of politics' and vice versa. This requires engaging with 'how material things, technologies and settings themselves become invested with more or less explicit political and moral capacities . . . how objects acquire "powers of engagement" and how those powers of engagement are articulated, discussed and contested in the public domain' (Marres & Lezaun 2011: 495). In short, the focus of inquiry turns to the engagements that the enrolment of different kinds of sociomateriality into a milieu of climate governance affords and the ways in which they come to be invested with different kinds of political purpose (Marres 2013), or what Blok (2014: 43) terms 'climatic assemblies in-the-making'. Using the example of an eco-show home, Marres explains that we should be concerned here with how such assemblies are 'accomplished in a distributed way, as the material setting, social actors, technologies and the visiting researcher/theorist all contribute to' affecting the forms of political participation available and in turn constituting what it is that is 'public' about domestic space whilst at the same time being constituted through this form of political practice (Marres 2013: 435). This in turn suggests that what is of analytical interest are the ways in which climate affords different gatherings, not in the historical sense of political assembly but rather in terms of the assemblies that are bought into alignment and the forms of politics they sustain. This is not a linear or discrete process, but rather a topological and unfolding one such that it is possible to trace multiple forms of climate publics from the 'effects of proximity, distance, and authority arise from shifting and particular arrangements of objects and their displacements through space and time' (Blok 2014: 48). The geographies of climate's politics are then neither straightforwardly local nor global but emerge in the juxtapositions of topological relations which in turn enable or circumvent their different forms of circulation (Bulkeley 2005; Bulkeley et al. 2015).

From Climate Controversies to Climate Care

Opening up the question of climate's publics and its politics requires then an engagement not only with the ways in which practices of sociomaterial assemblage serve as the means through which authorisation is rendered, problematisation takes place and techniques of governing are deployed and undone (see Chapters 3–5), but also to how such processes of configuration and alignment come to be invested as matters of concern. The translation of particular conditions into an issue capable of generating concern has predominantly been seen as occurring when matters can no longer be contained. Much attention has been given to issues around which controversy arises: 'moments of ontological disturbance in which the things on which people rely as unexamined parts of the material fabric of their everyday lives become molten' (Whatmore 2013: 37). Such forms of excess are not, in themselves, sufficient, but rather 'for these matters of concern to become political issues . . . emergent concerned groups – say, a patient group or an environmental NGO – must undertake their own analyses, describe their ailments and make themselves heard' (Blok 2011: 456). It is also at the limit of the practice of government as the calculated effort to shape conduct that Li (2007a) locates the practice of politics as critique gains force and the compromises of governing can no longer be sustained.

In both senses, politics is regarded as a matter of that which cannot be contained and becomes in one way or another overtly controversial. Yet if attempts to govern and the 'witches brew' (Foucault 2000b: 232) that constitutes the practice of such programmes are instead regarded as inseparable, we might instead look for the practice of politics in the ongoing constitution and reconfiguration of the practice of government itself and the mundane as well as strategic ways in which this is accomplished. The making of matters of concern need then not only arise where they come to overflow governmental programmes but rather be part of the ongoing means through which particular sociomaterial assemblages come to be constituted and considered in climate terms. For concern is first and foremost a matter of recognition, of coming to recognise some 'men and things' as of 'importance, interest' in relation to climate change, a set of things that 'affects or touches' one and that implies relation of belonging, involvement, engagement, implication and care (*Oxford English Dictionary Online* 2014). The ways in which sociomaterial conditions come to be seen as matters of concern goes beyond only those around which controversy arises to encompass those that seem to afford ways in which some kind of *care* for climate change can be registered. In order to explore these dynamics of controversy and care in the making of climate publics, the rest of this section examines two of the case studies in turn. In Hexham, initial consensus about the need to address climate change at the community level through the development of new forms of renewable energy became controversial as the sociomaterial assemblage of hydro power became entangled with the resource politics of aquatic environments and fishing economies. Community energy as an assembly-in-the-making came to be undone by the multitude of different ends it was intended to serve and the unwieldy gathering this created. In Brixton, in contrast, the consensus that climate can be governed through other means, and that other issues in turn can be mobilised by climate, has led to a productive gathering of momentum as new alignments have fostered the circulation of finance and technologies that can afford to serve multiple purposes. Rather than this being on the one hand a case of the political and the other a case of the post-political nature of climate change, the analysis of how things come to be of climate concern shows that politics is apparent within both controversial and consensual matters.

Hexham River Hydro and Its Discontents

In Hexham, the emergence of the community-led hydro-power scheme to generate electricity from the River Tyne emerged at the confluence of multiple concerns. In the local community, the transnational Transition Towns network had established a group engaged with matters of climate change and peak oil, issues of fuel poverty were of growing concern as energy costs continued to rise and renewable energy technologies were being mobilised. The aims of Hexham River Hydro were thus encompassing in their scope:

> To develop a hydro-electric generation scheme thus securing long-term generation of renewable electricity from the River Tyne at Hexham. To use the hydro-electric generation facility as an educational and an environmental awareness raising tool, both for residents of the area and visitors to Hexham. To demonstrate that generating renewable electricity is compatible with protection and even enhancement of the amenity value of a major river. To attract more visitors to Hexham (hydro-power schemes – even small ones – have proved to be popular visitor attractions elsewhere) (Hexham Community Partnership 2011: 1).

To utilise the resulting income to secure long-term funding for the low carbon and regenerative activities of HCP (Hexham Community Partnership 2011: 1).

For the partners who established the EnergyShare scheme, which provided both momentum and funding to the development of Hexham River Hydro–British Gas and River Cottage, the matters of concern related both to understanding the potential of different kinds of energy technologies and to the ways in which communities might come to be involved, implicated or engaged in the provision and use of different kinds of energy. The translation of these conditions into the specific configuration of Hexham River Hydro was orchestrated through the governing of a host of sociomaterial entities towards these conjoint but diverse finalities related through the idea of seeking to improvethe climate. Diffuse and encompassing, the will to govern climate change served as a rationale around and through which new junctures could be drawn and alignments created. As discussed previously, the practice of governing entails not only techniques designed to authorise, delimit, calculate and commensurate (Chapters 4 and 5), but also the bringing together of a public capable of forming attachment to climate-as-hydro: practices required for establishing a climate-assembly-in-the-making (Blok 2014). In this regard, the EnergyShare scheme was catalytic. What had been an embryonic idea championed by a few individuals found a means through which it could be made to matter. Once selected as a finalist, the scheme had to attract sufficient public support to win an online, time-limited public vote. The scheme had one month to gather its public in a process the organisers likened to voting on a popular TV programme:

> So it was a public vote, I always say it's a bit like *Strictly Come Dancing,* you know, it's that kind of mentality, you've got to show your feathers and folk vote for you and that's before you've actually done all the work, you win the money to do the work to do the feasibility work (Hexham River Hydro Interview B 2012).

Unlike the TV programme, with Hexham River Hydro there was in this case nothing tangible to vote for, nothing to see or admire except the area on the River Tyne where the hydro scheme, if built, would be situated. Generating public concern required other means as the multiple networks that cut through the hydro scheme's diverse assemblage were mobilised. This involved investing Hexham River Hydro with a set of attachments to its local place and at the same time to imagined global futures. Media opportunities and photo calls (involving local MPs, the MEP, councillors and representatives from community organisations in Hexham) used an inflatable globe with the message 'Support HRH – Save Tomorrow Today', and multiple reports appeared in the media (*Journal* 2011a, 2011b). At the same time, Hexham River Hydro sought to embed itself within the community through discussions with local schools and attendance at workshops on 'green' issues that a range of private, public and voluntary groups attended (HRH 2011). It was this aspect of making Hexham River Hydro a matter of local concern that organisers felt had enabled it to compete with much larger towns that had 'access to bigger populations' (Hexham River Hydro Interview B 2012):

> I think one of the reasons why it did get, you know, actually more votes than anybody else, it actually became an advantage, this small market town, because you are in touch with many more networks and many more people actually know what it's about and feel some ownership of it, I mean, that must have been the case because we did secure the biggest number of votes (Hexham River Hydro Interview B 2012).

Yet the making of Hexham River Hydro's public was not only as a local matter but as an issue that concerned those aligned to the Transition Towns movement and for whom Hexham provided some form of attachment – either as a site through which to think of 'home' or as a place where climate could come to be engaged:

> And we got quite a lot of people that were in touch with relatives, you know, that had a link with Hexham and had moved away or something, and then we also used the *Hexham Courant*, Hexham's Internet paper, you know, so there's a lot of ex-pats from Hexham that receive that, there's also a mail version as well, a mailed-out version too, which says, you know, something about Hexham readers even abroad, and we got votes from there too (Hexham River Hydro Interview B 2012).
>
> We used the international transition network as well e-mailing people to say, for example, in Canada, you know, if you feel like voting for us and if you've got any link to Hexham, we're in this competition, we'd like your support and we got two or three folks and two or three transition groups e-mail us and say, oh certainly, yes, I've voted and I'll ask all our members too (Hexham River Hydro Interview B 2012).

The climate-assembly-in-the-making that constitutes Hexham River Hydro has a topological politics, linking both the proximate and the distant, the present and the future, into a set of conditions that are seen to matter not, at least initially, because of their controversial nature or because they exceed the parameters of what can be governed, but precisely because they afford the generative possibilities of enacting and becoming implicated with the will to improve climate change.

In bringing this assembly into alignment, Hexham River Hydro encompasses a great many things and their varied sociomaterial composition. Mobilising the will to improve climate change, like other governmental programmes, has a tendency to be expansive (see Chapter 5). As Foucault explains, 'the apparatuses of security [or as he later described it, the programmes of government] . . . have a constant tendency to expand; they are centrifugal. New elements are constantly being integrated. . . . Security therefore involves organising, or anyway allowing the development, of ever-wider circuits' (Foucault 2009: 45). Hexham River Hydro is no exception. As its public is assembled, the matters with which Hexham River Hydro is concerned are continually reworked and reconfigured. The concentration of greenhouse gas emissions in the atmosphere intermingle with local housing conditions, energy costs, public finances, government subsidies for renewables, land ownership and engineering designs. As the scheme gathers momentum and begins to be calculated as a programme for intervention, it unsettles existing alignments between things and their publics, calling into question taken-for-granted orders and the normative capacities of sociomaterial configurations. Nowhere is this more apparent than in relation to Hexham River Hydro's very stuff – the river and the hydro-power installation.

As Hexham River Hydro gathered momentum, it acquired new kinds of practice intended to calculate how hydro power might be designed and installed in the river and necessarily open up the assemblage to intervention and contestation. Consultations are required. Affected groups were listed in a database. Meetings were held to discuss the project and its intentions. At the public meeting held in March 2012, the local MP, Guy Opperman, who chaired the meeting, was reported as stating that:

> It will be up to the Hexham Hydro team to convince the community, the anglers, the riparian owners and the County Council and Environment Agency that a hydro can be added without compromising the river and the fish who swim in it. That is a big hurdle for them to overcome. No one wants anything that will in any way damage the river and what runs through it (*Journal* 2012a).

In the case of the River Tyne, whilst different constellations of 'the river and what runs through it' made their presence felt, in the form of swimmers, rowers and canoeists and their objections to the ways in which hydro power might reconfigure the utility of the watercourse, it was the potential risks to its salmon ecology around which the making of Hexham River Hydro came to turn. The River Tyne has a reputation as the 'best salmon fishing river in England and Wales' (*Journal* 2011c) and there is a dedicated website – Fish Tyne – which is a wholly owned subsidiary of the Tyne Rivers Trust (TRT), which looks after 'the interests of the Tyne Catchment' (Fish Tyne 2014). Every micro-hydro project is charged with assessing the impacts and potential risks for fish species and taking steps such as the design of the scheme and the inclusion of protective screens and new/upgraded fish passes. In so doing, the governing of micro-hydro as a response to climate change draws into its compass a host of socio-material entities around which assemblies and forms of public order are already established. The Tyne Rivers Trust responsibility for the River Tyne (and in particular its fish stocks) predates the Hexham River Hydro project. Indeed, The Tyne Rivers Trust had already exposed problems for migratory fish at the Hexham Bridge site as a result of the mining of gravels in the 1950s and had conducted a feasibility study to replace the existing fish pass with a new one (*Journal* 2011c). Once the Hexham River Hydro scheme won the EnergyShare competition, the Tyne Rivers Trust issued an early position statement on its concern for the river's ecological and economic assets (a diverse range of migratory and coarse fish). The micro-hydro project should go ahead, the Trust argues, if there is 'certainty that it will not damage the river, its fish stocks and associated wildlife'. Whilst not against hydro technology per se, the Trust does expect that calculations of potential impacts be scientific:

> The Trust is not opposed to the principle of the hydro schemes on the river generally, or at this location. However, we believe that any hydro proposal must demonstrate, with clear scientific evidence, that it will not cause any physical damage to fish and other wildlife, or interfere in any way with the natural migration process (for example by noise, distraction, obstruction or reduced water flows). We have yet to see such evidence, or indeed a full Environmental Impact Survey, in relation to the proposal for Hexham Bridge. Similarly, we have not yet seen any proposals for the monitoring of any scheme, should it be built, or for its removal in the event that practice demonstrates that it causes unforeseen harm. Until we have seen such evidence and satisfactory proposals, we will continue to urge the relevant regulators to exercise caution when considering any formal applications for this, or any other hydro scheme in the Tyne Catchment (Tyne Rivers Trust 2012).

The Rivers Trust, the umbrella body under which the Tyne Rivers Trust operates, similarly positioned Hexham River Hydro interventions as raising new matters of concern that went beyond climate change and the production of renewable energy to encompass river system habitats and fishery economies. Drawing these things into relation means that there are no 'win-wins with hydro power, there's only a compromise, it's not the same thing' (Hexham River Hydro Interview C 2012). Whilst the insertion of the Hexham River Hydro scheme seeks to put the river to new purpose, its existing assemblies are not so easily displaced. Rather, the Hexham River Hydro scheme comes to be further integrated into the river's existing cohabitations as the design and location of the fish pass are negotiated in order to broker a compromise with TRT and its ambitions for the improvement of the Tyne's salmon fishery. However, just as the matters of concern appeared to have stabilised in this manner, a more organised and higher-profile objection from the local angling community arose. With the movement of the Hexham River Hydro project to the planning approval stage, the angling community sought to reopen the question of the impact of the hydro project on the fish pass:

> Insufficient monitoring has been carried out to date on the fish passage at Hexham Bridge, so there's no baseline information. Without that, we will not fully understand the implications of the scheme on fish passage, and how to mitigate it if necessary. We're not against hydro generation or green energy developments, but we feel that if the Tyne is to continue to be England's finest salmon river, full assessments and monitoring at Hexham Bridge must be carried out before a planning application is submitted (Andrew Jackson, honorary secretary of the Hexham Anglers Association, quoted in *Journal* 2012b).

The angling network is well organized and includes the Angling Trust and its legal arm, Fish Legal (Fish Legal 2014). The Angling Trust has explicitly developed a campaign aimed to 'Fight the Hydropower' (Angling Trust 2013), using the potential threat of climate change to river ecologies to destabilise the notion that hydro power might be able to 'improve' climate change:

> We care passionately about climate change, but low-head hydropower will make a negligible contribution to reducing carbon dioxide emissions and will damage river systems for everyone, not just anglers! (Angling Trust 2013).

Rather than being fixed, climate's publics, and their discontents, come to be realised through the process of accomplishing climate governance.

Micro-hydro, a relatively unknown technology, has to be made resonant with a range of communities – it has to be made to matter. Accomplishing micro-hydro as a public requires that it is practiced in relation to a whole host of entities to which it might form attachments. At the same time, this expansive circulation encompasses things that cannot so readily be accommodated, and existing assemblies seek to provide alternative accounts of with what climate governance should, and should not, be concerned.

Made in Brixton: Climate Care

In Brixton, the pursuit of the governing of climate change took multiple guises. During the 2000s, as London and its mayor increasingly promoted the city as a space for climate change responses (see Chapters 2 and 3), local authority planning, social landlord building refurbishment and the grassroots activities of the Transition Town group all referenced the need to reduce the use of fossil fuel-based energy in order to respond to climate change. It was the introduction in 2009 of the Low Carbon Zone that served as one means through which these different interventions came to be gathered and assembled in relation to one another. Promoted by Boris Johnson in his first two years as mayor, the Low Carbon Zones programme was intended to develop a prominent suite of initiatives engaging municipal authorities, private sector interests and the public at large and formed part of the attempt to make the London 2012 Olympics 'the greenest games ever' (BBC News 2012). Brixton Low Carbon Zone, one of ten successful projects funded over the period 2009–12, was a clearly delimited area of Lambeth Borough:

> The zone contains 721 buildings and extends north from the centre of Brixton to Max Roach Park and east from the commercial centre of Brixton Road to incorporate the Canterbury Gardens, Loughborough and Hertford estates. . . . This area has been chosen because without intensive carbon reduction activity Brixton cannot achieve its goal of becoming a green hub. The purpose of the Brixton LCZ is to address these challenges and secure Brixton's future as a leader in carbon reduction (London Borough of Lambeth 2009a: 2).

This clearly circumscribed boundary intentionally drew a cordon around those matters that were to be of climate concern and the publics that should necessarily be engaged and those that lay beyond. Yet, as with any boundary device, its demarcation belies its attachments to matters that exceed the zone and through which the zone itself is constituted. Rather than being determined institutionally or socially, the boundaries of the Low Carbon Zone were designed to capture particular sociomaterial assemblages that were regarded as having potential to animate its governing intentions:

> The way the project was designed was to bring in a lot of the land that has its own capital funding . . . so the plan was that that money would be used for leveraging in external funding. So we've just got confirmation that there's another programme, it's called Community Energy Savings programme . . . and we have got a good offer through Eon to do a major programme for the Loughborough estate. So that will help deliver the retrofit programme and the measures like external/internal wall insulation and cavity fills, loft insulation, boiler replacements, improved windows to a decent home standard . . . there was a time when we weren't sure Lambeth Council would be in a position to allocate money towards it because of the whole financial restrictions. And that would have been a serious concern in terms of us achieving our target. But we're really happy that Lambeth has been able to prioritise this and the main reason being that we were able to leverage in funding from a utility company like Eon. So they're bringing in three million pounds – so that added value. But then also . . . the other reason that they chose to deal with us, because obviously it would bring them up to standard, the homes – we would be able to address fuel poverty, we would be able to provide decent homes for a large number of residents (Brixton Interview A 2011).

As with the case of Hexham River Hydro, the governmental programme realised through Brixton's Low Carbon Zone is centrifugal, drawing into its compass a range of new elements. Climate comes to be coupled here with an existing range of sociomaterial concerns both for its own sake – to enable climate change to become compelling – and as a means of addressing matters that appeared dormant and in need of revitalisation. Climate here acts as a catalyst to investing these sociomaterial assemblages with new purposes, and in turn these provide a means of making climate a matter of concern:

> . . . it is around community cohesion. It is around addressing fuel poverty, bringing homes to a decent standard, you know, it is about reducing carbon emissions because that – in terms of the GLA funding that is the main aim: the project aims to bring about reduction of 20.12 per cent in time for March 2012, for the Olympics, basically. That is overarching but . . . you know Brixton Low Carbon Zone is a bit more about a holistic approach because we don't think that if you just pitch it at, you know, all we hear about energy and carbon reduction, that people would engage in the whole process (Brixton Interview A 2011).

Such concerns were not, though, self-evident. Rather, accomplishing climate governance through the Low Carbon Zone required fostering attachments to the zone, its projects and practices. For those participating in this endeavour, the notion of community provided a fundamental means through which such connections should be forged. Yet, as an essentially contested concept, what community entailed and for whom it could be established provoked a significant challenge. To start with, there was a great deal of musing over what community meant in a diverse and economically divided place such as Brixton, where multiple ethnic communities, communities related to specific housing estates and those orchestrated by particular interests or issues could all be identified. Jostling for position amongst these notions of community, the Low Carbon Zone assembly found little traction not least because those involved 'found it hard to actually communicate the Low Carbon Zone as a community because it's very arbitrary' (Brixton Interview B 2011). Rather, the assembly of the public for whom the Low Carbon Zone might come to matter had to be made, woven in practice through the recurrent practice of crafting connections between people and things. In a rather tautological manner, then, addressing community was regarded as the way into the community:

> We knew that for it to be effective if you just used normal general communications it wouldn't really penetrate through the community, we wouldn't be able to get them on board. By investing in people skills and looking at the community engagement aspect of things, is the only way we could address a lot of the barriers and get the key messages through, so... addressing the community cohesion aspect of things would be the way into the community (Brixton Interview A 2011).

Fostering this engagement between communities and climate change required more than general communications but instead the grafting of low carbon in and through existing forms of assembly. The diverse finalities with which the practice of the will to govern climate is concerned enabled it to weave together existing interventions and new finances through a host of sociomaterial configurations, from the growing of local food to ensuring comfortable living spaces. One important means through which this was achieved was through the Green Doctors project. Partly managed under the auspices of the LCZ, this six-month scheme was funded by Groundwork via the Future Jobs Fund, providing placements for five people who were trained not only to identify the improvements required but to offer 'treatment' to those sociomaterial conditions that could be improved in relation to climate change:

> It wasn't funded by Low Carbon Zone, it was part of our bid. We did it and we said having Green Doctors, going and installing ten easy measures and draught-proofing within the zone. And so we did use some of the GLA money, quite a bit in the first year of funding, we used quite a bit in terms of purchasing draught strips and radius panels, energy monitors and energy efficient lightbulbs so that when our Green Doctors were ready, then they could install these measures in the households, talk to people about energy, address a lot of energy management issues, behaviour change and all of that (Brixton Interview A 2011).

Between May and October 2010, the team of Green Doctors completed 150 home energy visits that included providing energy advice and installing energy-saving devices such as energy-saving light bulbs, radiator panels and electricity monitors. As Marres and Lezaun (2011: 495) suggest, such ordinary devices when deployed in particular settings can be 'carefully designed and arranged to produce particular effects and facilitate the investment of those objects with participatory potential.' The suite of mundane things gathered and aligned around notions of saving energy, carbon and the climate, invested with the specific purpose of building connections and forging alliances capable of revitalising domestic space and its importance in relation to wider social and environmental goals, is not merely a matter of reshaping material flows of energy and money or of the practicing of a particular technique. Rather, such installations serve as a means through which what it means to be affected with energy and climate comes to be established. Here, climate's public and its politics are not determined by the specific physical properties belonging to any one device, but rather its materialisation is 'the result of an operation afforded by the device, that of the de- and re-composition of everyday material action, an operation that codes this action in particular terms' (Marres 2011: 517), in this case that seemingly small changes can be made to make a difference.

Such materialisations of climate's public were not confined to individual households or domestic space, but rather served as an important means through which new connections could, quite literally, be woven together:

> A lot of feedback and a lot of the advice that the Green Doctors were giving is that you can address a lot of the issues by having thicker curtains and lining curtains ... doable, simple measures that everyone can do themselves. And it

just came off of that, okay, there's a really high demand for this and then we ran curtain lining workshops, and people did actually come and use the facility. So we're going to do that again just before winter sets in properly, we'll run some more of that, because we have lots of materials. So it's just the little projects like that that really ... get people more engaged in the whole process. And all we've done is been able to provide them the space. We book the space, got the tables, got the sewing machines for them and had someone to help them do it (Brixton Interview A 2011).

So we were doing our Green Community Champions. ... And running draught-busting workshops, and we got a few more people involved in running draught busting, and they were like, 'oh, I'd really like to know how you do this', and so the Transition Town people were taking it on. And now, today, they're a social enterprise and set up on their own and they are known as Community Draught Busters. And we helped them in terms of putting a business model together (Brixton Interview B 2011).

Here, investing in the materialities of curtain making workshops or draught-busting enterprises, and the capacities for addressing climate change and energy poverty that they could afford, served as the means through which the conditions gathered through the Low Carbon Zone could be translated into forms of climate assembly. Such capacities are not given by the material politics involved in assembling, aligning and installing new things but rather is a set of 'acquired capacities' generated relationally through the topological connections that link mundane household devices with urban politics and global environmental issues (Marres 2013).

This is a politics devoid of controversy but woven through issues of concern. Being concerned with the condition local housing, the affordability of energy bills and the carbon implications of energy consumption emerges not as the practice of government gives way to a practice of politics. Rather, it is central to the ways in which the practice of government is continually being made political through the assembly of the sociomaterial publics through which it is realised. Such climate-assemblies-in-the-making (Blok 2014) come to be invested with concern, which in turn provide ongoing momentum to the ways in which governing is accomplished. In Brixton, investment took literal form through the development of Brixton Energy, 'a not-for-profit co-operative ... [which creates] cooperatively owned renewable energy projects whose financial revenues stay within the local community' (Brixton Energy 2014). Initiated through the Low Carbon Zone and Transition Town Brixton's energy group, the cooperative works by offering shares in the investment in solar polyvoltaic (PV) panels, which are then installed on social housing in the community:

> Brixton Energy Solar 1 saw the installation of a 37kW solar array on the roof of Elmore House on the Loughborough Estate. The array has been generating renewable energy for Brixton since 30th March 2012. Solar 2 saw the installation of a 45kW solar array on Styles Gardens – five housing blocks situated on the Loughborough Estate. The array has been generating renewable energy for Brixton since the 31st of October 2012. To date, the two projects have generated in excess of 50,000kWh of community-owned renewable energy in Brixton (Repowering London 2014).

A third installation is currently under offer. Starting at a pound a share, the schemes were deliberately designed to make energy, PV and climate change matters in which all manner of households could invest and through which the risks of producing local power could be shared. Climate's public here was concerned with building collective power to challenge existing systems of energy provision and enable the generation of energy to create a return on the investments made by the community. Seeking to redirect the financial flows of the energy economy to those least able to afford energy, the project entailed a radical and progressive politics. Yet this political moment emerges not through radical forms of agonistic encounter, but rather through mundane materials and the practice of governing by technique that have often been regarded as in some sense post-political (Swyngedouw 2010). This analysis suggests that rather than politics being a matter of confrontation, it is a matter of attachment – of how, by whom and with what implications attachments come to be made to climate change, and how climate itself fosters publics capable of realising an array of outcomes that matter.

Conclusions

That climate change is an issue of concern is rather taken for granted amongst those who seek to address or analyse its politics. Thinking through the ways in which issues are constituted and matters come to be of concern with the literature on material politics and the material constitution of publics casts things in a different light. Issues are made to matter, and their materiality and the forms of engagement this does (and does not) afford are critical in this respect. This in turn means that climate's politics are not confined to particular times and spaces, such as the decision-making process of local planning committees or the annual Conference of the Parties, which is heralded with so much fanfare, but rather are performed 'in settings and through objects that do not belong to a distinct sphere of action, but rather co-articulate public political activity with other domains of everyday practice' (Marres & Lezaun 2011: 496). Formed in the practice of governing, making matters of concern and the political assemblies through which they come to register, 'requires hard work of symbolic and material articulation' (Blok 2011: 470). Accomplishing climate governance both produces and is produced through the making of such climate assemblies. The cases here show how such assemblies are not formed on virgin territory, but are a matter of bricolage, as practices of authorisation, comprehension and articulation weave together and repurpose things and their politics through a mesh of topological networks.

Such perspectives give a new and vital insight into the whereabouts of climate governance, its politics and its geographies. To date, this work has focused primarily on how and why such issues become controversial, exceeding the boundaries of existing political space and acquiring new meaning and power. Keeping politics alive has been seen as a matter of ensuring its ongoing contestation, of creating space for the vitality of matter, its unfolding and unanticipated consequences, on the one hand, and for the (verbal) articulation of political demands on the other. Controversy is essential to climate's politics. As the case of Hexham River Hydro showed, once climate change comes to be animated with particular things, the matters that concern, the issues that are raised and the assemblies that are formed can open up the ground for considering and contesting how, why and to what ends climate should be governed in very explicit terms. For many working in the field, it is this boundary of where governing stops and politics starts that is of critical interest. Issues become political at the point of failure or of transcendence: where the success (or constriction) of governing comes undone and issues are made properly political in that they are opened up to their own contradictions, and alternatives can be voiced in an agonistic manner.

Yet this chapter has also found that things come to matter through different registers and forms of relating. One of the most remarkable things about the politics of climate change is how uncontroversial it can be. Such a politics of consensus (see Chapter 3) can be viewed as serving the interests of powerful groups that channel the problem of climate change in particular ways to serve their ends and at the same time enrolling a host of less powerful players towards these goals. There is no doubt evidence of these forms of politics as climate capitalism unfolds and the interests of business and political elites become attached to those of addressing climate change (Paterson & Newell 2010). Yet at the same time, climate consensus can be formed not through relations of exploitation but through those of care and, perhaps, solidarity. In Brixton, climate change became a means through which to assemble a host of issues of concern, from fuel poverty to vegetable growing, around new forms of care. Here, climate change is carefully positioned to articulate what it is that matters in particular places and to draw into articulation the kinds of resources and networks that enable new assemblies to be formed and a new form of politics to emerge. Climate change has then no particular politics. Its politics are made by the ways in which climate governance is accomplished.

7
Conclusion

Introduction

The stampede to buy powerful vacuum cleaners intensified yesterday as retailers reported a fourfold surge in sales. Shoppers made their feelings clear about an EU ban on models of more than 1,600 watts by visiting stores or online sellers in their droves. Many have been stockpiling two or more high-power models to beat regulations that came into effect yesterday. The new EU directive is intended to reduce energy use, but it has infuriated the public, who say they'll have to spend longer using the appliances. (Daily Mail 2014)

James Dyson has suggested that Britain should leave the EU, over a dispute about vacuum cleaner energy efficiency regulations. A new law banning vacuum cleaners rated above 1,600 watts came into effect on September 1, as part of the European Commission's plans to meet targets on energy efficiency. Dyson was not affected by the changes, as none of its vacuum cleaners are rated above 1,400 watts. However, the company's founder claims the legislation is too crude and does not go far enough. (Telegraph, 2014)

That climate change is a political matter has been clear ever since its emergence in scientific and public discourse more than two decades ago. Yet our understanding of *the ways in which* it is political has remained relatively fixed on questions of the actors, institutions and mechanisms through which a global commons could or should be managed. There is limited space in such an account for considering how the politics of climate change come to matter in a diverse range of sites and arenas – to consider its multiple political geographies. Take for example the row that blew up in the British press in the autumn of 2014 surrounding the realisation that new European legislation would lead to the banning of certain types of vacuum cleaners with the most powerful engines, in the name of energy efficiency and (in turn) climate protection. Vacuum cleaners are not the most obvious site for arguments about how society should respond to climate change or indeed whether the United Kingdom should remain part of the European Union. Yet both issues came to be entangled as commentators took sides seeking to support the beleaguered British housewife seeking to keep her home free from dust on the one hand and the innovative capacities of British industry to support the development of a green economy on the other. Here, climate's politics came to be forged not through the agreement of timetables and targets for managing the global atmosphere but through the design, manufacture, standardisation, regulation and contestation of both what it meant to vacuum and how, why and through whom we might want to conserve energy.

It is with these alternative geographies of the politics of climate change that this book is concerned. To address climate politics in this way has entailed a focus on the questions of how climate governance is accomplished: by what means, where, for whom and with what implications. Attending to climate politics as a matter of governing is first and foremost to acknowledge that power comes in various guises. As violence, as control, as domination, power is also mobilised in and through climate change. Yet, for the time being at least, such forms of power remain marginal. In contrast, forms of governing – authorised practices of rule and conduct – proliferate. For all the attention that has been directed to these forms of climate politics, our understanding of how governing is accomplished, of how it is fulfilled and performed and of the ways in which its pursuit require the development and deployment of skills and technique remains limited. Examining the ways in which governing is *accomplished*, this book has sought to open up questions of the political geographies of climate governance to critical interrogation. Starting from a perspective that conceives of power in facilitative and immanent terms and of governing as a programmatic and sociomaterial process, it has sought to explore the ways in which climate change serves to foster particular spatial orderings (e.g., as community, as business, as city), how these are contended and in turn how climate change is constituted through the topologies of power through which such ways of bringing the world to order are made. Rather than occupying the fixed territorial demarcations of the global, national, regional and local, an analysis of the ways in which climate governance is accomplished also illuminates the new forms of governance space that are configured and contested as particular socio-material assemblages are bought into alignment and held together through comprehending climate change.

Once the matter of the governing of climate change is opened up in this way, its tendency to get everywhere – from the corridors of power to the power of vacuum cleaners – is less surprising than it might at first appear. Rather than being secured, tied to particular issues, domains or places, governing is a programmatic activity that continually unfolds: a constantly evolving suite of practices and processes intended to intervene in particular sociomaterial conditions accompanied by the promise of their improvement (Li 2007a; Rose & Miller 1990). The nature of governing and the pursuit of diverse ends suggest in turn that it has a 'constant tendency to expand' (Foucault 2009: 45). What Foucault terms at this point in his writings the 'apparatuses of security', what we might also consider to be the workings of particular forms of governmental intervention or assemblage, are 'centrifugal. New elements are constantly being integrated' (Foucault 2009: 45). In other words, accomplishing climate governance is an ever-expanding activity whereby more and more diverse elements are integrated into the programme of improving the condition of climate change. Rather than expecting climate governance to remain contained within one or other political arena or governance space, we should instead seek to trace the ways in which new elements are enrolled into the circulations of climate governance and the consequences both for what it means to govern climate change and to undertake anything from shopping to vacuuming, EU negotiations to the design of new technologies.

In seeking to explore the means through which climate governance is accomplished through these circulations and topologies of power, the book explored the dynamics of authorising, ordering and articulating climate change and the ways in which it was made public in a diverse range of cases drawn from the United Kingdom. The United Kingdom, as outlined in Chapter 2, has been an intriguing place through which to examine the ways in which climate politics unfold. Both heralded and lambasted for its state-led responses to the climate agenda, the United Kingdom has also been a fertile ground for many alternative sites and arenas of climate politics, from the business of boardrooms to the working of community energy projects. Whilst one of the aims of the book has been to provide an overview of the multiple facets of climate politics in the United Kingdom, its purpose has not been to evaluate or integrate these phenomena in order to provide a comprehensive picture of the United Kingdom's response to climate change and the level or extent of its effectiveness. Rather, the examples drawn upon have served as lenses through which to explore the dynamics by which accomplishing climate governance is pursued. The aim in this concluding chapter is similarly to draw out the insights that these cases provide into how and with what implications the governing of climate change is taking place.

Beyond the Climate Problem

As the climate problem has come to be known as one of atmospheric concentrations of greenhouse gases, predictions of changing metrological conditions, fragile ecosystems and social vulnerabilities, a parade of various authorities has come into view seeking to delimit what it is to which we must attend and how this could be achieved. Scientific authority has been central to this endeavour, but so too has the personal authority of politicians and movie stars; the economic authority of big business lined up for or against action; and the moral authority of religious leaders, community groups and so on and so forth. Yet, as important in accomplishing climate governance are the often hidden practices through which authorisation is generated and granted. Rather than being preordained, this account suggests that authorisation is both immanent to the working of power and generative in its constitution. In short, that authorisation is not pre-given but achieved in and through the working of governance (see Chapter 3). From this perspective, there is no one authoritative position, actor or institution and no final accepted authority upon which to base climate governance, but rather authorisation is constituted through the practices and processes of governing climate change. This in turn suggests that climate change is not governed at a distance, as overarching principles or rationalities for appropriate social responses are relayed to communities and individuals, but rather the authority to govern is made through the forging of multiple proximities which in turn serve to constitute both the domain to be governed and its spatial and temporal horizons. In turn, as authorisation is produced, what constitutes the climate problem comes to be reworked and reconfigured in relation to those entities who are party to its assemblage and to the compromises required to bring it into alignment as a comprehendible and ordered domain for intervention.

Examining the ways in which state-led programmes for community governance of energy and climate change have emerged in the United Kingdom, practices of consent and acquiescence were found to be commonplace. Securing finance has required that community groups of various guises concerned with a spectrum of issues around questions of sustainability, energy and poverty ensure that their activities deliver in relation to United Kingdom government ambitions for emissions reductions. More critically, in seeking a community basis for the delivery of emissions targets, government itself is concerned with the more than metric, with what community-based responses can generate in terms of the immeasurable qualities of engagement and well-being. Into this space of auditing benefit and consenting to the goals of state-based programmes have emerged a range of intermediary actors, creating new forms of relation between communities, the state, carbon (markets), finance and various material interventions, from renewable energy to loft insulation. These forms of intermediation have fostered consent whilst at the same time supporting the diverse finalities with which community-based initiatives are concerned and the forms of improvement that such programmes seek to realise alongside carbon reduction. This was not a matter of designated authority or of the authority of particular institutions or individuals, but of the relational generation of authorisation in practice.

Whilst authorisation as consent certainly has a part to play in the ways in which climate governance is accomplished, by far its most dominant mode comes in the guise of consensus. Although the early years of climate politics were notable for a marked antagonism concerning what an appropriate response to the climate problematic might involve, particularly between those entities with fossil fuel–based interests and those with less explicit ties to such resources, since the turn of the twenty-first century such forms of open disagreement have receded from view in the United Kingdom and Europe (though they remain alive and well in other contexts, such as the United States, Australia and South Africa). For some, the emergence of climate consensus is central to a post-political condition in which forms of properly political contestation are evacuated from the site of politics. Yet the analysis of the practice of authorisation as consensus in the cases here suggests that they are far from compliant with dominant orderings of climate change, but rather serve as means through which fragile alignments of climate change with other entities are generated whilst at the same time being subject to ongoing contestation and reconfiguration. Consensus *is a more than discursive process* through which an array of sociomaterial conditions come to be made to fit in with one another and in turn become implicated in how climate change comes to be comprised. Consensus is a binding activity, serving to delimit the bounds of what constitutes climate governance, to share the risk and responsibility for drawing such lines. This in turn requires the work of fitting climate change into existing orders and entraining a multitude of entities into climate change such that what it means to govern in this domain becomes enmeshed in a host of practices and processes.

The work of authorisation is therefore not an abstract matter, but one generated through assembling and aligning a host of different entities with the problematisation of climate change. That governing is a problematizing activity 'linked to the problems around which it circulates, the failings it seeks to rectify, the ills it seeks to cure' (Rose & Miller 1992: 181), is now well rehearsed. The very notion of governing as the orchestration of the will to improve (Li 2007a) and Foucault's notion of government as a mode of power exercised through the conduct of conduct in order to improve the condition of the population (2009: 105) in relation to 'a series of specific finalities' (2009: 99) suggest that it is with the 'problems' of the population, as an amalgam of 'men and things' (2006: 96), that government is concerned. The cases of the Hexham River Hydro scheme and the practices of making climate change count at Tesco supermarkets illustrated the ways in which the governing of climate change required the gathering together of diverse elements to assemble what might (and what might not) constitute the domain through which climate change could be addressed. Whilst this involved the translation of the problem of climate change, as one of energy consumption, efficiency and alternative forms of production, into the spaces of a local community and a corporate body, it also involved the bringing into climate change of a range of diverse elements from production chains to local housing. Such work was only partly centred on and through particular problems, of the need to address energy poverty, improve business efficiency, create financial resources – of attending to the diagnosis of problems. Instead, this was work that addressed the notion of improvement as a matter of opportunity, of the *diagnoses of possibility* that climate change could afford. Here, climate governance is accomplished through attending to the desires and multiple ends it helps to foster and realise.

That rendering climate change as a matter amenable to governance involves not only processes of problematisation but also diagnoses of possibility points the analysis of its politics towards the ways in which it comes to be comprehended – understood, encapsulated and embraced. To date, most work in the field of climate politics that has adopted a post-structural reading of power, mirroring wider debates on governmentality and the social economy of markets, has focused on the technical means through which climate change has come to be known, on matters of statistics, models, audits, accounting, measurement and calculation of various kinds. No doubt these are vital means through which what climate change is come to be understood. Yet a focus only on those things that can be counted, or rendered technical, limits how an issue such as climate change comes to be comprehended beyond the calculative. In focusing on how climate change is made intelligible, the ways in which climate *sensibilities* are constituted and contested can be obscured. In Brixton, for example, the demarcation of the Low Carbon Zone was at once an act of calculation, of drawing a zone on a map that would encompass the kinds of subjects that could be governed, be they buildings or people, in line with the objectives of carbon reduction. Yet at the same time, it was the sensibility of the zone –its capacity to feel climate change, the dynamic potential of the ways in which climate change could be connected to emergent agendas and existing forms of intervention – that shaped the how climate governance was accomplished. At HSBC, making climate change governable was even more explicitly tied to the ways in which the issue came to be assembled through personal encounters that worked through ensuring not only that climate change was visible, but that it was felt and experienced firsthand. Bearing witness to the ways in which climate change could matter, HSBC sought to engage the body corporate with a sensibility towards climate change through which it became ordered as a matter through and around which the corporation could and should be concerned.

From this analysis, it is clear that bringing climate change to order entails a set of practices and processes that go beyond the climate problem in at least three important ways. First, the politics of climate change exceeds what is commonly thought of as the climate problem. Making climate change governable requires the bringing into alignment of a diverse array of entities that are more or less figured in climate terms. In this process, climate is translated into multiple finalities, from energy security to corporate responsibility, the making of markets to new infrastructure projects, the development of new products or the retrofitting of housing, whilst multiple other objects and subjects are bought into relation with climate change. In this sense, the boundaries of what constitutes climate change are not given but are immanent to the ways in which the will to improve is orchestrated and assembled. Second, the politics of climate change go beyond the climate *problem* in that they also attend to its *possibilities*. The notion of problematisation is necessary but not sufficient to capture the ways in which climate governance is accomplished. Rather, we need to attend to the ways in which possibilities, desires, hopes and futures are mobilised in relation to governing climate change and the ways in which this serves in turn to draw into alignment a diverse range of entities through which climate change comes to be enacted. Finally, the analysis suggests that accomplishing climate governance goes beyond the problem as conventionally understood through techniques and practices that seek to render the field of intervention intelligible, known and in some sense measured to also encompass the means through which climate change sensibilities are forged and mobilised. Together, this suggests that the governing of climate change is not only or perhaps not most productively thought about as a clearly circumscribed sphere for intervention, but an unfolding set of processes and dynamics through which what it means to govern climate change are always contested and realigned.

More Than the Sum of the Parts

Accomplishing climate governance presents endless possibilities. The expansive and circulatory nature of governmental power combined with the number of distinct finalities to which the systemic challenge of the ways in which society produces and uses energy can be attached means that almost anything could be rendered in climate change terms. Yet the landscape of climate governance is far from complete, with some sociomaterial conditions bought into alignment and others remaining outside its comprehension. Central to understanding how some things come to be assembled in climate governance terms whilst others do not are the techniques and practices through which governing is conducted. In the work on climate governance, much attention has been directed towards two particular kinds of techniques, those of calculation and commensuration, through which climate change comes to be rendered technical, abstracted, joined together and managed through various centres, notably those of the associated with the carbon economy. As Chapter 5 discusses, much of the analysis of these practices of government has sought to examine how they work to enable government at a distance and the rolling out of neoliberal logics through different kinds of governance space. The perspectives developed in this account of climate's politics suggest that any ready interpretation of the work of calculation and commensuration as taking place at a distance needs to be tempered with an understanding of the ways in which such techniques work in practice through the continual translation and incorporation of techniques and devices into the sociomaterial order and the means through which diverse forms of proximity are achieved. In short, that rather than being a matter of rolling out governmental techniques or imposing them from a distance, it is the ways and means by which they are articulated that is central to how climate governance is accomplished.

Seeking to understand the practice of calculation and commensuration in these terms requires that they are seen not only as abstractions but also as forms of material intervention in the configuration of governmental assemblages (Lansing 2012: 207). How the value of carbon is calculated, or offsets are accounted, serves not only a representational function but also composes the field of which it is a part. It serves to make some parts of the sociomaterial order more amenable to intervention and to exclude others. As material interventions that are highly contested, calculation and commensuration are unstable and contingent devices whose outcomes are always provisional and that require continual work and maintenance (Lovell 2014). As Callon (2009: 542) argues, 'an issue always finds an exit enabling it to overflow'. In terms of climate politics, the devices through which it comes to be calculated and made commensurate are seen as able only to capture and contain part of the issue, allowing it to overflow beyond the spaces, such as carbon markets, through which its governance is designated. These arguments about the continual interplay between the devices of government and the matters to which they seek to attend have been central in shaping the perspective developed here. Yet the cases suggest that there are other dynamics at work in the practices of calculation and commensuration that are also important in understanding the ways in which they serve to enable the governing of climate change. Critically, they suggest that such techniques are not only practices of *abstraction* but are also orchestrated through *embedding* intentions and desires, drawing diverse elements into one another and fostering new articulations and ways of comprehending different sociomaterial orders. In short, calculation and commensuration work in a dual sense, both abstracting from the entanglements that go to make up particular assemblages and render them amendable to governmental intervention and at the same time serving to configure and embed particular assemblages, to dispose them towards particular ends.

Perhaps rather curiously, much of the analysis of calculation as a form of governmental technique has focused on either single instruments or domains through which it is deployed. The analysis here starts instead from examining calculation as a practice, looking at the multiple ways in which it is performed as climate change comes to be comprehended and governmental programmes enacted. It suggests that rather than being a singular activity or one-off intervention, calculation is a multiple and ongoing process. The case of the Berwick CoRE wind turbine, for example, showed how forms of calculation beget other calculations, and how this practice moves between registers from mechanical engineering to bank loans. In one sense, this requires a degree of abstraction, serving to derive the potential of wind energy and translate it across these different arenas. At the same time, it becomes means through which these entanglements are forged and the will to improve climate governance is fostered through distinct sociomaterial articulations. It is then the interweaving of calculation, its repetition and refraction across and through particular sociomaterial orders, that serves to enable climate governance to be accomplished. This analysis also revealed a rather intriguing phenomenon. Rather than seeking to escape the configuration of climate governance, the cases across the book illustrate the *inflows* of entities and instances that seek to be counted in climate change terms. The drawing into articulation of this multiplicity is part of the important work served by the practices of calculation, which in turn relies not only on measurement but also on diagnoses of the value that is accrued by more than the sum of the parts.

Likewise, analysing the practice of commensuration, it is clear that what is at stake is not so much 'making things the same', in the way that market devices work through the standardisation of value, but of finding what, from Fabiana Li (2011), we can call a *logic of equivalence.* Rather than abstracting value, this in turn requires that a host of entities are bought into alignment and puzzled together into new articulations. In the case of Tesco, making climate change fit into the body corporate involved a host of interventions, from the design of stores to the calculation of carbon, the fitting of new fridge doors to the work routines of employees. Some of this work is implicit, as Tesco sought to make its intention to govern carbon work on the ground whilst explicitly rendering new blueprints for store design and making the carbon content of products visible were intended to signal this new order and enable it to circulate. Indeed, it is in enabling the circulation of climate change that the practices of calculation and commensuration are so vital. Given that governing has a centripetal tendency, an expansive logic that seeks to enable the circulation across the population of the will to improve, it is perhaps unsurprising that such techniques enable this to be performed. Rather than this being a matter of abstraction, however, this analysis suggests that it is through articulation, the coming together of things and their multiple connections, that such forms of circulation are enabled, in turn allowing climate change to be articulated, spoken of and known, beyond the specific sites of intervention. The opposite is also true: the misfit or incommensurability of entities in the articulation of climate change governance serves to limit the ability of any intervention to hold things in proximity and enable circulation. Projects fail, programmes fail into disrepair and the comprehension of climate change fades.

Accomplishing climate governance requires then more than the sum of the parts – more than the sum of particular metrics, more than the sum of diverse forms of calculation and more than the sum of making things the same. It is instead a matter of how things come to be regarded as worthy of equivalence and connected with one another on at least some form of common ground. The inflow of entities into the domain of climate governance serves to animate and reconfigure what it is that accomplishing climate governance entails, in turn requiring new practices of authorisation and comprehension. These practices of articulation, comprehension and authorisation work together in establishing the field to be governed and the ways in which this can be achieved, such that what it means to govern climate change and how it comes to matter are continually being reconceived.

A Matter of Concord

Accomplishing climate governance serves not only to rearticulate sociomaterial orders towards particular ends but at the same time to make them matters of public concern, to make them political. Considering politics in material terms and of publics wrested from distinct institutional settings and instead in relation to those things that are of concern suggests that the sites and moments of climate's politics are indeed diverse and far-ranging. In the field of science and technology studies, much attention has been directed to the importance of acknowledging the ways in which the formation of publics takes place in relation to material assemblages and interventions that come to be of concern (Braun and Whatmore 2010; Latour 2005). This in turn means that political participation is 'performed in the workplace, the home, nature, the market – in settings and through objects that do not belong to a distinct sphere of action, but rather co-articulate public political activity with other domains of everyday practice' (Marres and Lezaun 2011: 496). Rather than climate's publics existing in predetermined arenas or specific domains, such as the international negotiations or community-based organisations, such perspectives suggest that they are formed through the means by which its government is accomplished. For the most part, this literature has attended to the ways in which such publics are formed as matters of concern overflow or come to break loose from governmental programmes, creating controversies that are not readily accommodated by existing sociomaterial orders. Politics in this sense is seen as encompassing those sociomaterial relations that emerge at the edge or limit point of governance.

This need not be the case. Making matters of concern can also be regarded as part of the ongoing ways in which climate governance is accomplished as particular assemblages come to be authorised, comprehended and articulated in climate terms. Politics is in this reading made through and with governance. In the case of Hexham River Hydro, for example, a relatively unfamiliar technology has to be made present and relevant to its context and the range of communities through which it is articulated as a response to climate (and other) issues. In short, it has to become a matter of attachment, of concern. However, the encompassing and circulatory nature of climate governance draws into the fold multiple entities whose fit and attachment are not straightforward. Kept separate, salmon fishing and the provision of renewable energy for a community are relatively uncontroversial. Articulated with one another and in relation to climate change, the things that come to matter are reconfigured, and what does and does not fit is laid bare. Accomplishing climate governance entails in such instances addressing the disturbances that the inflow of multiple entities entails, of managing the population, the assemblage, 'in depth, in all its fine points and details' (Foucault 2009: 107).

The work of accomplishing climate governance entails, then, alongside authorisation, comprehension and articulation, the means by which to practice concern. Controversy is perhaps its most visible and readily available form. On the one hand, this is managed through consent, such as those practised through multiple forms of assessment, from environmental audits to financial checks, by which controversial elements are bought into equivalence to test their fit and recognise the limits of what can and cannot be included in climate governance. On the other hand, it is managed through protest and resistance, as entities resist being included and subject to governmental programmes trained on climate change objectives. Here one overflow might lead to another, as for example in the Hexham River Hydro case the economic value of producing hydro power comes to be articulated with that of producing a salmon fishing river, other concerns about value and investment also begin to circulate, and the forms of improvement that hold the intervention together as an assemblage begin to unravel. This in turn leads to new articulations of what it means to govern climate change as its spatial and temporal orderings are recast.

The cases in this book suggest that controversy is not the only means through which concern for climate change is articulated and performed. Indeed, many of the means through which climate change comes to matter are rather mundane, unobtrusive and tangential. A host of ordinary things, from the inhabitants of woodlands outside of Oxford that are measured in climate change terms, to the contents of the weekly shop to the making of curtains provide the things through which climate care is expressed and performed. Accomplishing climate governance entails the coupling of climate change with multiple other entities, both as the intention to improve the condition of climate change expands and circulates and as other entities seeking investment are corralled towards the climate change domain. This is in a sense the work of consensus, of establishing the common ground of climate change as a matter around which a host of concerns can be articulated and jointly realised, whilst at the same time acknowledging the compromises that this entails in terms of the goals and intentions of other programmes. Attending to the material politics of these processes reveals too that what is at stake is also the formation of public concord on climate change. As a practice of authorisation, concord relies on 'the "bringing into agreement" of heterogeneous elements – social and material – in such a manner that they are taken for granted as authoritative in the shaping of conduct' (Bulkeley 2012: 2439). As political participation takes place outside of designated spaces of political authority, its practice entails forms of conduct in line with making particular things matter that may be taken for granted or routinized in relation to what constitutes appropriate or good conduct – the putting of materials in the right bins, for example. The cases here suggest that achieving such forms of concord, so that climate care is authorised in everyday practice, is a complex and precarious achievement. In Brixton, for instance, things, from solar panels to draught-busting workshops, serve as an important means through which climate care is 'bought into agreement' with the practice of everyday life, whether this be in particular moments of proximity or through the routinisation of new ways of acting in the world. It matters then not only by whom, where, how and why climate governance is accomplished, but also through which sociomaterial means.

Climate Resolve?

For over two decades, climate change has been positioned as a problem for which society needs to find a solution. Much of the ambition to solve climate change has been invested in the international community, the seemingly logical space through which to find a common approach to tackle our common concerns. Whilst a growing body of evidence amassed through the field of global environmental governance has suggested that there are other ways and means through which climate change might be addressed, many critiques suggest that they will be insufficient for solving the climate problem. The questions that have been at the heart of understanding climate's governance have followed suit, inquiring into the effectiveness, efficiency and sustainability of interventions; how they might be made more useful or more complete; and how we might attend to the issues of legitimacy and justice that they raise. These are all important issues, not least because of the investment placed in seeking climate solutions from the international negotiations to experiments in geoengineering that have a host of potential social and environmental issues. At the same time, the contention here is that they are also incomplete. What is more, they have served to position the governing of climate change as involving or requiring particular sets of authority, agency and power that are, on the one hand, capable of solving the problem and yet do not fully capture its potential and its politics. Rather than seeking to position climate change as a problem of this kind and attending to the ways in which it can be solved, the arguments advanced here have sought to explore the ways in which climate makes things political and in turn how the working of power and politics makes climate change. Taken together, it suggests that climate governance is accomplished through authorisation, comprehension, articulation and the constitution of material publics. The politics and geographies of governing climate change are to be found woven through these practices and formed in their configuration, connecting seemingly diverse elements – such as the vacuum cleaners with which this chapter started – to climate change.

Rather than seeing climate change as a problem, this analysis suggests we might more productively consider it to be a *condition*, and in Foucault's sense, one that relates to the population, where the population is considered to comprise 'men in their relationships, bonds and complex involvements with things' (2009: 96). As a condition, climate change cannot be solved in the traditional sense (Hulme 2009: 361–2). Whilst we may be able to make major efforts to reduce atmospheric concentrations of greenhouse gases and ameliorate the impacts of the changing climate, we cannot be rid of climate change as a condition. Much like other conditions around which society's aspirations accumulate, from peace to the rights of women, climate change comes to be both part of our socio-material order and a domain through which we seek the improvement of ourselves and others. This is a politics that emerges not in the space of the public, but in the (res)public. It requires then not that we solve climate change, but that we *re-solve* it. Climate resolve requires that we look anew at the sites and arenas through which it might come to be more progressively or productively configured. This book has sought to chart a range of such interventions and to examine what they tell us about how the governing of climate change can be accomplished. It demonstrates that climate resolve requires, as the multiplicity of definitions of the term suggests, that it is at once decomposed and extended across the socio-material body, converted into other things and orchestrated such that it is able to transform 'dissonant harmony ... to form a concord' (*Oxford English Dictionary Online*). Gathering the resolve to address climate change requires then the marshalling of intention, of the will to improve the climate condition, but at the same time of enabling its circulation and extension into the multiple domains and sites by which it comes to be authorised, comprehended, articulated and made to matter. Rather than investing so much of our climate resolve in the international domain, we can instead further its circulation through the geographically dispersed and yet connected sites and arenas through which climate governance is already taking place. This will entail becoming much more accomplished at these forms of governing. It may also be a means through which we can accomplish much more to improve the climate condition towards more progressive ends.

References

Abbott, Kenneth (2012) The transnational regime complex for climate change, *Environment and Planning C: Government and Policy*, 88 (3): 543–64.

Agnew, John (2005) Sovereignty regimes: Territoriality and state authority in contemporary world politics, *Annals of the Association of American Geographers*, 95 (2): 437–61.

Agnew, John (2011) Waterpower: Politics and the geography of water provision, *Annals of the Association of American Geographers*, 101 (3): 463–76.

Agyeman, Julian and Evans, Bob (1994) The new environmental agenda, in Agyeman, J. and Evans, B. (eds) *Local Environmental Policies and Strategies*, Longman, Harlow, pp. 1–22.

Allen, John (2003) *Lost Geographies of Power*, Blackwell, Oxford.

Allen, John (2004) The whereabouts of power: Politics, government and space, *Geografiska Annaler* 86 B (1): 19–32.

Allen, John (2010) Powerful city networks: More than connections, less than domination and control, *Urban Studies*, 47(13) 2895–911.

Allen, John (2011) Topological twists: Power's shifting geographies, *Dialogues in Human Geography*, 2011 (1): 283–98.

Allen, John and Cochrane, Allan (2010) Assemblages of state power: Topological shifts in the organization of government and politics, *Antipode*, 42 (5): 1071–89.

Angling Trust (2013) Fighting hydropower, available online: www.anglingtrust.net/page.asp?section=496&sectionTitle=Fight+the+Hydropower (accessed February 2015).

Armstrong, Andrea and Bulkeley, Harriet (2014) Micro-hydro politics: Producing and contesting community energy in the North of England, *Geoforum*, 56: 66–76.

Barnett, Michael and Duvall, Raymond (2005) Power in international politics, *International Organization*, 59 (1): 39–75.

Barry, Andrew (2001) *Political Machines: Governing a Technological Society*, Athalone Press, London.

Barry, Andrew (2013) *Material Politics: Disputes along the Pipeline*, Wiley Blackwell, Chichester, UK.

BBC News (2012) London 2012: Olympic Games 'greenest ever', 11 August 2012, available online: www.bbc.co.uk/news/uk-19220847 (accessed February 2015).

Bernstein, Stephen (2011) Legitimacy in intergovernmental and non-state global governance, *Review of International Political Economy*, 18 (1): 17–51.

Berwick Advertiser (2012) Funding is secured for town wind turbine, 30 May 2012, available online: www.berwick-advertiser.co.uk/news/local-headlines/funding-is-secured-for-town-wind-turbine-1-2326564 (accessed February 2015).

Berwick Trust (2014) Wind Turbine Project, available online: http://berwicktrust.org.uk/?page_id=15 (accessed February 2015).

Betsill, Michele M. and Bulkeley, Harriet (2006) Cities and the multilevel governance of global climate change, *Global Governance* 12 (2): 141–59.

Biermann, Frank and Dingwerth, Klaus (2004) Global environmental change and the nation state, *Global Environmental Politics* 4 (1): 1–22.

Biermann, Frank and Pattberg, Philipp (2008) Global environmental governance: taking stock, moving forward, *Annual Review of Environment and Resources* 33: 277–94.

Blok, Anders (2011) Clash of the eco-sciences: Carbon marketization, environmental NGOs and performativity as politics, *Economy and Society*, 40 (3): 451–76.

Blok, Anders (2014) Experimenting on climate governmentality with Actor-Network Theory, in Stripple, J. and Bulkeley, H. (eds) *Governing the Climate: New Approaches to Rationality, Power and Politics*, Cambridge University Press, Cambridge, pp. 42–58.

Boykoff, Maxwell and Goodman, Michael (2009) Conspicuous redemption? Reflections on the promises and perils of the 'celebritization' of climate change, *Geoforum*, 40 (3): 395–406.

Brace, Catherine and Geoghegan, Hilary (2010) Human geographies of climate change: Landscape, temporality, and lay knowledges, *Progress in Human Geography* 35 (3): 284–302.

Braun, Bruce and Whatmore, Sarah (eds) (2010) *Political Matter: Technology, Democracy, and Public Life*, University of Minnesota Press, Minneapolis and London.

British Gas (2012) Welcome to Green Streets, available online: www.britishgas.co.uk/smarter-living/save-energy/green-streets.html (accessed February 2014).

Brixton Energy (2014) Brixton Energy: Power by, to and for the people!, available online: https://brixtonenergy.co.uk/ (accessed February 2015).

Brook Lyndhurst and Ecometrica (2011) Review of the climate challenge fund. Edinburgh, Scottish Government. Available online: www.scotland.gov.uk/Resource/Doc/352709/0118663.pdf (accessed 4 March 2013).

Bulkeley, Harriet (2001) Governing climate change: the politics of risk society?, *Transactions of the Institute of British Geographers*, 26 (4): 430–47.

Bulkeley, Harriet (2005) Reconfiguring environmental governance: Towards a politics of scales and networks, *Political Geography*, 24: 875–902.

Bulkeley, Harriet (2012) Governance and the geography of authority: Modalities of authorisation and the transnational governing of climate change, *Environment and Planning A*, 44 (10): 2428–44.

Bulkeley. Harriet, Andonva, Lilliana, Betsill, Michele M., Compagnon, Daniel, Hale, Tom, Hoffmann, Matt, Newell, Pete, Paterson, Mat, Roger, Charlie and VanDeveer, Stacy (2014) *Transnational Climate Change Governance*, Cambridge University Press, New York.

Bulkeley, Harriet and Betsill, Michele M. (2003) *Cities and Climate Change: Urban Sustainability and Global Environmental Governance*, Routledge, London.

Bulkeley, Harriet and Betsill, Michele M. (2013) Revisiting the urban politics of climate change, *Environmental Politics*, 22 (1): 136–54.

Bulkeley, Harriet, Castán Broto, Vanesa and Edwards, Gareth (2015) *An Urban Politics of Climate Change: Experimenting with Socio-technical Transitions*, Routledge: London.

Bulkeley, Harriet, Castán Broto, Vanesa and Maassen, Anne C. (2014) Urban infrastructure networks and the governing of climate change, *Urban Studies*, 51 (7): 1471–86.

Bulkeley, Harriet and Kern, Kristine (2006) Local government and climate change governance in the UK and Germany, *Urban Studies*, 43 (12): 2237–59.

Bulkeley, Harriet and Newell, Peter J (2015) *Governing Climate Change*, 2nd edition, Routledge, London.

Bulkeley, Harriet and Schroeder, Heike (2008) Governing climate change post-2012: The role of global cities – London. Tyndall Working Paper No. 123, University of East Anglia.

Bulkeley, Harriet and Schroeder, Heike (2012) Beyond state/non-state divides: Global cities and the governing of climate change, *European Journal of International Relations* 18 (4): 743–66.

Bulkeley, Harriet, Watson, Matthew and Hudson, Ray (2007) Modes of governing municipal waste, *Environment and Planning A*, 39 (11): 2733–53.

Callon, Michele (2009) Civilizing markets: Carbon trading between in vitro and in vivo experiments, *Accounting, Organizations and Society* 42 (3–4): 535–48.

Callon, Michele, Millo, Yuval, and Muniesa Fabian (eds) (2007) *Market Devices*, Blackwell, Oxford.

Cardiff Transition (2014) What is transition?, available online: http://cardifftransition.com/about (accessed December 2014).

Carlsson, Sverke, Paterson, Matthew and Stripple, Johannes (2005) Privatising governance, practising triage: Securitization of insurance risks and the politics of global warming, in Levy, D. and Newell, P. (eds) *The Business of Global Environmental Governance*, MIT Press, Cambridge, MA, pp. 249–274.

Carter, Neil (2008) Confronting climate change in the UK: Challenges and obstacles, *Political Quarterly*, 79 (2): 194–205.

Carter, Neil (2014) The politics of climate change in the UK, *Wiley Interdisciplinary Reviews: Climate Change* 5 (3): 423–33.

Carter, Neil and Jacobs, Michael (2014)) Explaining radical policy change: The case of climate change and energy policy under the British Labour government 2006–2010, *Public Administration* 92 (1): 125–41.

Cashore, Ben (2004) Legitimacy and the privatization of environmental governance: How non-state market-driven governance systems gain rule-making authority, *Governance*, 15 (4): 503–29.

Castree, Noel (2002) Environmental issues: From policy to political economy, *Progress in Human Geography* 26 (3): 357–65.

Castree, Noel, Adams, William M., Barry, John, Brockington, Daniel, Büscher, Bram, Corbera, Esteve, Demeritt, David, Duffy, Rosaleen, Felt, Ulrike, Neves, Katja, Newell, Peter, Pellizzoni, Luigi, Rigby, Kate, Robbins, Paul, Robin, Libby, Rose, Deborah Bird, Ross, Andrew, Schlosberg, David, Sörlin, Sverker, West, Paige, Whitehead, Mark and Wynne, Brian (2014) Changing the intellectual climate, *Nature Climate Change*, 4 (9): 763–68.

Catney, Philip, MacGregor, Sherilyn, Dobson, Andrew, Hall, Sarah Marie, Royston, Sarah, Robinson, Zoe, Ormerod, Mark and Ross, Simon (2014) Big society, little justice? Community renewable energy and the politics of localism, *Local Environment: The International Journal of Justice and Sustainability*, 19 (7): 715–30.

CBI [Confederation of British Industry] (2014) The colour of growth: Maximising the potential of green business. Available online: www.cbi.org.uk/media/1552876/energy_climatechangerpt_web.pdf (accessed February 2015).

Climate Group, The (2007) Blair and Schwarzenegger meet business leaders to accelerate climate change solutions. Available online: www.theclimategroup.org/what-we-do/news-and-blogs/Blair-and-Schwarzenegger-meet-business-leaders-to-accelerate-climate-change-solutions/ (accessed February 2015).

Climate Group, The (2012) The Climate Group Partnership wins top business charity award. Available online: www.theclimategroup.org/what-we-do/news-and-blogs/the-climate-group-partnership-wins-top-business-charity-award/ (accessed February 2015).

Climate Group, The (2014a) Our values. Available online: www.theclimategroup.org/who-we-are/our-values/ (accessed February 2015).

Climate Group, The (2014b) Our partners: HSBC Holdings PLC. Available online: www.theclimategroup.org/who-we-are/our-members/hsbc-holdings-plc (accessed February 2015).

Climate Group, The (2014c) HSBC Climate Partnership Celebration Event. Available online: www.theclimategroup.org/what-we-do/events/HSBC-Climate-Partnership-Celebration-Event (accessed February 2015).

Climate Group, The (2014d) 10 Years of the Climate Group. Available online: www.theclimategroup.org/our-achievements/ (accessed February 2015).

Climate Week (2014) Green your cuisine: Meat free Monday, available online: www.climateweek.com/green-your-cuisine/ (accessed February 2015).

Collier, Ute (1997) "Windfall" emissions reductions in the UK, in Collier, U. and Löfstedt, R. (eds) *Cases in Climate Policy: Political Reality in the European Union*, Earthscan, London, pp. 87–107.

Committee on Climate Change (2014) Meeting carbon budgets – 2014 progress report to Parliament Committee on Climate Change. Available online: www.theccc.org.uk/wp-content/uploads/2014/07/CCC-Progress-Report-2014_web_2.pdf (accessed February 2015).

Conca, Ken (2005) Old states in new bottles? The hybridization of authority in global environmental governance, in Barry, John and Eckersley, Robyn (eds) *The State and the Global Ecological Crisis*, MIT Press, Cambridge, MA, pp. 181–206.

Core Cities Group (2007) A prosperous and sustainable future. Available online: www.corecities.com/sites/default/files/images/publications/a%20prosperous%20and%20sustainable%20future.pdf (accessed February 2015).

Creamer, Emily (2014) The double-edged sword of grant funding: A study of community-led climate change initiatives in remote rural Scotland, *Local Environment: The International Journal of Justice and Sustainability*. DOI: 10.1080/13549839.2014.885937.

Daily Mail (2014) Power surge! Fourfold rise in sales of super vacuums: Some customers buying two or more models to beat new EU regulations, 2 September. Available online: www.dailymail.co.uk/news/article-2740197/Power-surge-Fourfold-rise-sales-super-vacuums-Some-customers-buying-two-models-beat-new-EU-regulations.html#ixzz3SYyDMWkF (accessed February 2015).

Dean, Mitchell (1999) *Governmentality: Power and Rule in Modern Society*, London: Sage.

Dean, Mitchell (2007) *Governing Societies: Political Perspectives on Domestic and International Rule*, Open University Press, Maidenhead.

Death, Carl (2014) The limits of climate governmentality, in Stripple, J. and Bulkeley, H. (eds) *Governing the Climate: New Approaches to Rationality, Power and Politics*, Cambridge University Press, Cambridge, pp. 77–91.

DECC [Department for Energy and Climate Change] (2012) Low carbon communities challenge evaluation report. Available online: www.gov.uk/government/uploads/system/uploads/attachment_data/file/48458/5788-low-carbon-communities-challenge-evaluation-report.pdf (accessed February 2015).

DECC [Department for Energy and Climate Change] (2014a) Reducing the UK's greenhouse gas emissions by 80% by 2050. Available online: www.gov.uk/government/policies/reducing-the-uk-s-greenhouse-gas-emissions-by-80-by-2050/supporting-pages/eu-emissions-trading-system-eu-ets (accessed February 2015).

DECC [Department for Energy and Climate Change] (2014b) Community energy strategy. Available online: www.gov.uk/government/uploads/system/uploads/attachment_data/file/275169/.20140126Community_Energy_Strategy.pdf (accessed February 2015).

DEFRA [Department for Environment, Food and Rural Affairs] (2012) Rural Communities Energy Fund: Main points. Available online: www.gov.uk/government/uploads/system/uploads/attachment_data/file/183280/rcref-mainpoints.pdf (accessed February 2015).

Department of the Environment (1990) *This Common Inheritance*, London: HMSO.

Department of the Environment (1994) *Climate Change: The UK Programme* London: HMSO.

Descheneau, Patrick and Paterson, Matthew (2011) Between desire and routine: Assembling environment and finance in carbon markets, *Antipode*, 43 (4): 662–81.

DETR [Department for Environment, Transport and the Regions] (2000) Climate change: The UK programme. Available online: http://webarchive.nationalarchives.gov.uk/20110523172013/www.decc.gov.uk/en/content/cms/what_we_do/change_energy/tackling_clima/programme/programme.aspx (accessed February 2015).

DTI [Department for Trade and Industry] (2003) Our energy future: Creating a low carbon economy. Energy White Paper. Available online: http://webarchive.nationalarchives.gov.uk/20090609003228/www.berr.gov.uk/files/file10719.pdf (accessed February 2015).

DTI [Department for Trade and Industry] (2006) The energy challenge: Energy review report 2006. Available online: webarchive.nationalarchives.gov.uk/20070603164510/www.dti.gov.uk/files/file31890.pdf (accessed February 2015).

DTI [Department for Trade and Industry] (2007) Meeting the energy challenge: An energy white paper. Available online: http://webarchive.nationalarchives.gov.uk/20121205174605/www.decc.gov.uk/assets/decc/publications/white_paper_07/file39387.pdf (accessed February 2015).

Dowling, Robyn (2010) Geographies of identity: Climate change, governmentality and activism, *Progress in Human Geography* 34 (4): 488–95.

Earthwatch (2013a) Interview with Rachel Phillips, Head of Learning, Earthwatch. Accessed online: www.earthwatch.org/europe/hcp/interview_rphillips/ (accessed March 2013, no longer available).

Earthwatch (2013b) HSBC Climate Partnership. Accessed online: www.earthwatch.org/europe/hcp (accessed March 2013, no longer available).

Earthwatch (2013c) Citizen science. Accessed online: www.earthwatch.org/europe/newsroom/corporate_partnerships/news-6-iccb.html (accessed March 2013, no longer available).

Energyshare (2011) Hexham River Hydro blog. Accessed online: http://groups.energyshare.com/hrh/blogs/hrh-supporters-enjoyed-the-autumn-sunshine/ (accessed February 2015).

Espeland, Wendy and Stevens, Mitchell (1998) Commensuration as a social process, *Annual Review of Sociology*, 24: 313–43.

Eyre, Nick (2001) Carbon reduction in the real world: How the UK will surpass its Kyoto obligations, *Climate Policy*, 1: 309–26.

Feola, Giuseppe and Nunes, Richard (2014) Success and failure of grassroots innovations for addressing climate change: The case of the Transition Movement, *Global Environmental Change* 24: 232–50.

Fernandez, Sara (2014) Much ado about minimum flows... Unpacking indicators to reveal water politics, *Geoforum*, 57: 258–71.

Finlayson, Alan (2011) 'What will yours grow into?' Governmentality, governance, social science and the Child Trust Fund, *Policy & Politics* 39 (4): 547–66.

Fish Legal (2014) Welcome to Fish Legal. Available online: www.fishlegal.net/landing.asp?section=158&sectionTitle=Fish%20Legal (accessed February 2015).

Fish Tyne (2014) About the Tyne. Available online: www.fishpal.com/England/Tyne/AboutTheRiver.asp?dom=Tyne (accessed February 2015).

Forum for the Future (2014) The big shift. Available online: www.forumforthefuture.org/thebigshift (accessed February 2015).

Foucault, Michael (2000a) The subject and power, in Faubion, J. (ed) *Power: Essential Works of Foucault 1954–1984*, Volume 3, New Press, New York, pp. 326–48.

Foucault, Michael (2000b) Questions of method, in Faubion, J. (ed) *Power: Essential Works of Foucault 1954–1984*, Volume 3, New Press, New York, pp. 223–38.

Foucault, Michael (2009) *Security, Territory, Population: Lectures at the College de France 1977–1978*, ed. Senellart, M., trans. Burchell, G., Palgrave Macmillan, Basingstoke.

Foxon, Timothy J., Hammond, Geoffrey P., Pearson and Peter J. G. (2010) Developing transition pathways for a low carbon electricity system in the UK, *Technological Forecasting & Social Change* 77: 1203–13.

Freidberg, Susanne (2013) Calculating sustainability in supply chain capitalism, *Economy and Society*, 42 (4): 571–96.

Friedrich, Carl J. (1958). Authority, reason and discretion, in Friedrich, C. J. (ed.), *Authority*, Harvard University Press, Cambridge, MA.

Friends of the Earth (2014) History of the big ask. Available online: www.foe.co.uk/news/big_ask_history_15798 (accessed February 2015).

Fuller, Sara and Bulkeley, Harriet (2013) Energy justice and the low carbon transition: Assessing low carbon community programmes in the UK, in Bickerstaff, K., Walker, G. and Bulkeley, H. (eds.) *Energy Justice in a Changing Climate*, Zed, London, pp. 61–78.

Gabriel, Nate (2014) Urban political ecology: Environmental imaginary, governance, and the non-human, *Geography Compass*, 8 (1): 38–48.

Geuss, Raymond (2008) *Philosophy and Real Politics*.,Princeton University Press, Princeton, NJ.

Ghertner, D. Asher (2010) Calculating without numbers: Aesthetic governmentality in Delhi's slums, *Economy and Society*, 39 (2): 185–217.

Gibbs, David, Longhurst, James and Braithwaite, Clare (1998) 'Struggling with sustainability': Weak and strong interpretations of sustainable development within local authority policy, *Environment and Planning A*, 30: 1351–65.

Gouldson, Andy and Sullivan, Rory (2012) Long-term corporate climate change targets: What could they deliver? *Environmental Science and Policy*, 27: 1–10.

Greater London Authority (2004) *The London Plan: Spatial Development Strategy for Greater London*, GLA, London.

Greater London Authority (2007) *Action Today to Protect Tomorrow: The Mayor's Climate Change Action Plan*. GLA, London.

Greater London Authority (2008) *The London Plan: Spatial Development Strategy for Greater London (Consolidated with Alterations since 2004)*, GLA, London.

Greater London Authority (2009a) Mayor announces £3m to kickstart 10 'energy-busting' low carbon zones in the capital, Greater London Authority Press Release, 1 May.

Greater London Authority (2009b) *Leading to a Greener London: An Environment Programme for the Capital*, GLA, London.

Greater London Authority (2011a) *Delivering London's Energy Future: The Mayor's Climate Change Mitigation and Energy Strategy*, GLA, London.

Greater London Authority (2011b). Project review: RE:CONNECT (Low Carbon Zones), interim review presented at GLA Investment and Performance Board, 13 December.

Grocer, The (2012) 'Frustrated' Tesco ditches eco-labels, 28 January. Available online: www.thegrocer.co.uk/channels/supermarkets/tesco/frustrated-tesco-ditches-eco-labels/225502.article (accessed February 2015).

Guardian, The (2006) The year the world woke up, 20 December, p. 7.

Guardian, The (2012) Tesco drops carbon-label pledge, 30 January. Available online: www.theguardian.com/environment/2012/jan/30/tesco-drops-carbon-labelling (accessed February 2015).

Guardian, The (2014a) George Osborne defeated in attempt to weaken UK carbon budget, 22 July. Available online: www.theguardian.com/environment/2014/jul/22/george-osborne-defeated-in-attempt-to-weaken-uk-carbon-budget (accessed February 2015).

Guardian, The (2014b) EU leaders agree to cut greenhouse gas emissions by 40% by 2030, 24 October. Available online: www.theguardian.com/world/2014/oct/24/eu-leaders-agree-to-cut-greenhouse-gas-emissions-by-40-by-2030 (accessed February 2015).

Gunn, D. (2008) Climate challenge fund. Available online: www.sustainable-scotland.%20net/documents/4979_Climate%20Challenge%20Fund%2008.10.pdf (accessed 4 March 2013).

Gustavsson, Eva, Ingemar Elander and Mats Lundmark (2009) Multilevel governance, networking cities, and the geography of climate change mitigation: Two Swedish examples, *Environment and Planning C: Government and Policy*, 27 (1): 59–74.

Hajer M A, 1995 *The Politics of Environmental Discourse: Ecological Modernisation and the Policy Process*, Clarendon Press, Oxford.

Hajer, Maarten (2009) *Authoritative Governance: Policy-Making in the Age of Mediatisation*, Oxford University Press, Oxford.

Hale, Stephen (2010) The new politics of climate change: Why we are failing and how we will succeed, *Environmental Politics*, 19 (2): 255–75.

Hargreaves, Tom (2014) Smart meters and the governance of energy use in the household in Stripple, J. and Bulkeley, H. (eds) *Governing the Climate: New Approaches to Rationality, Power and Politics* Cambridge University Press: Cambridge, pp. 127.

Head, Lelsey and Gibson, Chris (2012) Becoming differently modern: Geographic contributions to a generative climate politics, *Progress in Human Geography* 36 (6): 699–714.

Hébert, Karen (2014) The matter of market devices: Economic transformation in a Southwest Alaskan salmon fishery, *Geoforum*, 53: 21–30.

Hexham Community Partnership (2011) Summary of feasibility study. Accessed online: www.hexhamcommunity.net/media/Microsoft%20Word%20-%20Summary%20HRH%20Feasibility%20Study%20July%202011%20v2.pdf (accessed September 2012, no longer available online).

Hexham Community Partnership (2013) Hexham River Hydro: Find out more. Accessed online: www.hexhamcommunity.net/pages/projects/hexham-river-hydro/hrh—find-out-more.php (accessed July 2013, webpage no longer available).

Hinchliffe, Steve (1996) Helping the Earth begins at home: The social construction of socio-environmental responsibilities, *Global Environmental Change*, 6 (1): 53–62.

HM Government (2009) The UK Low Carbon Transition Plan: National strategy for climate and energy. Available online: www.gov.uk/government/uploads/system/uploads/attachment_data/file/228752/9780108508394.pdf (accessed February 2015).

HM Government (2011) The carbon plan: Delivering our low carbon future. Available online: www.gov.uk/government/uploads/system/uploads/attachment_data/file/47613/3702-the-carbon-plan-delivering-our-low-carbon-future.pdf (accessed February 2015).

Hobson, Kersty and Niemeyer, Simon (2011) Public responses to climate change: The role of deliberation in building capacity for adaptive action, *Global Environmental Change*, 21 (3): 957–71.

Hodson, Mike and Marvin, Simon (2007) Understanding the role of the national exemplar in constructing 'strategic glurbanization', *International Journal of Urban and Regional Research*, 31: 303–25.

Hodson, Mike and Marvin, Simon (2009) 'Urban ecological security': A new urban paradigm? *International Journal of Urban and Regional Research*, 33 (1): 193–215.

Hodson, Mike and Marvin, Simon (2013) *Low Carbon Nation?* Routledge, London.

Hoffmann, Matthew (2011) *Climate Governance at the Crossroads: Experimenting with a Global Response after Kyoto*, Oxford: Oxford University Press.

HRH (2011) HRH supporters enjoyed the autumn sunshine, Energyshare HRH Blog, 17 November. Available online: www.energyshare.com/hrh/blogs/hrh-supporters-enjoyed-the-autumn-sunshine (accessed February 2015).

HSBC (2010) HSBC Climate Partnership Review 2010. Available online: www.theclimategroup.org/_assets/files/HSBC-Climate-Partnership-Review-2010_1.pdf (accessed February 2015).

HSBC (2013) HSBC Sustainability Report. Available online: www.hsbc.com/citizenship/sustainability/reports-and-documentation (accessed February 2015).

Hulme, Mike and Turnpenny, John (2004) Understanding and managing climate change: The UK experience, *Geographical Journal*, 170 (2): 105–15.

Hurd, Ian (1999) Legitimacy and authority in international politics, *International Organisation*, 53 (2): 379–408.

Ilcan, Suzan and Phillips, Lynne(2007) Developmentalities and calculative practices: The millennium development goals, *Antipode*, 42 (4): 844–74.

Jagers, Sverker. C. and Stripple, Johannes (2003) Climate governance beyond the state, *Global Governance*, 9: 385–99.

Jessop, Bob (2002) Liberalism, neoliberalism, and urban governance: A state-theoretical perspective, *Antipode* 34 (3): 452–72.

Jordan, Andrew and Hutimea, Dave (2014) Policy innovation in a changing climate: Sources, patterns and effects, *Global Environmental Change* 29 (2014) 387–94.

Journal, The (2011a) Gillian's dip has bought a warm glow, 3 December. Available online: www.thejournal.co.uk/news/local-news/gillians-dip-brought-warm-glow-4417791 (accessed February 2015).

Journal, The (2011b) Hexham Hydro team brave the cold in bid for £100,000 grant, 2 December. Available online: www.thejournal.co.uk/news/north-east-news/hexham-hydro-team-brave-cold-4417691 (accessed February 2015).

Journal, The (2011c) Hexham Road bridge weir is a threat to salmon, 1 July. Available online: www.thejournal.co.uk/news/north-east-news/hexham-road-bridge-weir-threat-4426733 (accessed February 2015).

Journal, The (2012a) Hexham Hydro meeting 'very positive' says MP, 30 March. Available online: www.thejournal.co.uk/news/north-east-news/hexham-hydro-meeting-very-positive-4413986 (accessed February 2015).

Journal, The (2012b) River groups urge delay in Hexham Hydro plans, 25 October. Available online: www.thejournal.co.uk/news/north-east-news/river-groups-urge-delay-hexham-4402640 (accessed February 2015).

Kenis, Anneleen and Mathijs, Erik (2014) Climate change and post-politics: Repoliticizing the present by imagining the future? *Geoforum* 52: 148–56.

Keohane, Robert. and Victor, David. (2011) The regime complex on climate change, *Perspectives on Politics* 9 (1): 7–23.

Knox-Hayes, Janelle (2013) The spatial and temporal dynamics of value in financialization: Analysis of the infrastructure of carbon markets, *Geoforum* 50: 117–28.

Kraftl, Peter (2014) Liveability and urban architectures: Mol(ecul)ar biopower and the 'becoming lively' of sustainable communities, *Environment and Planning D: Society and Space* 32: 274–92.

KSB [Keep Scotland Beautiful] (2013) Climate challenge fund application guidance notes. Available online: http://ccf.keepscotlandbeautiful.org/documents/CCFApplication%20GuidanceNotesJan20132660.pdf (accessed 4 March 2013).

Lansing, David M. (2012) Performing carbon's materiality: The production of carbon offsets and the framing of exchange, *Environment and Planning A*, 44: 204–20.

Lake, David (2010) Rightful rules: Authority, order and the foundations of global governance, *International Studies Quarterly*, 54: 587–613.

Latham, Robert. 1997. States, global markets and social sovereignty. Presented at SocialScience Research Council's Conference on Sovereignty and Security, Notre Dame, IN.

Latour, Bruno (2005) From Realpolitik to Dingpolitik: Or how to make things public, in Latour, B. and Weibel, P. (eds) *Making Things Public: Atmospheres of Democracy*, Exhibition at the KZM, MIT Press, Cambridge, MA, pp. 14–41.

Le Dantec, Christopher and DiSalvo, Carl (2013) Infrastructuring and the formation of publics in participatory design, *Social Studies of Science*, 43 (2): 241–64.

Leahy, Sir Terry (2009) SCI Conference: 16 October 2009 Royal Society, London Speech by Sir Terry Leahy, Tesco. Available online: www.tescoplc.com/index.asp?pageid=17&newsid=570 (accessed February 2015).

Lee, Kah-Wee (2014) Feeling like a state: Design guidelines and the legibility of 'urban experience' in Singapore, *International Journal of Urban and Regional Research*, 38 (1): 138–54.

Lemos, Maria Carmen and Agrawal, Arun (2006) Environmental governance, *Annual Review of Environmental Resources*, 31: 297–325.

Levy, David and Egan, Daniel (1998) Capital contests: National and transnational channels of corporate influence on the climate change negotiations, *Politics and Society*, 26: 337–61.

Levy, David and Newell, Peter (2005) *The Business of Global Environmental Governance*, MIT Press, Cambridge, MA.

Li, Fabiana (2011) Engineering responsibility: Environmental mitigation and the limits of commensuration in a Chilean mining project, *Focaal*, 60 (13): 61–73.

Li, Tania Murray (2005) Beyond 'the state' and failed schemes, *American Anthropologist*, 107 (3): 383–94.

Li, Tania Murray (2007a) *The Will to Improve: Governmentality, Development, and the Practice of Politics*, Duke University Press, Durham, NC.

Li, Tania Murray (2007b) Practices of assemblage and community forest management, *Economy and Society* 36 (2): 263–93.

Lipschutz, Ronnie (2005) Power, politics and global civil society, *Millennium* 33 (3): 747–69.

Litfin, Karen T. (2000) Environment, wealth, and authority: Global climate change and emerging modes of legitimation, *International Studies Review* 2 (2): 119–48.

Local Government Association (2014) Climate local. Available online: www.local.gov.uk/climate-local (accessed February 2015).

Loconto, Allison (2014) Assembling governance: The role of standards in the Tanzanian tea industry, *Journal of Cleaner Production*, online first: doi:10.1016/j.jclepro.2014.05.090.

London Borough of Lambeth (2009a) Application for Low Carbon Zone.

London Borough of Lambeth (2010) Low Carbon Zone briefing April 2010.

London First (2006) Draft further alterations to the London Plan response from London First to the mayor of London, July.

Lorenzoni, Irene, O'Riordan, Tim and Pidgeon, Nick (2008) Hot air and cold feet: The UK response to climate change, in Compston, H. and Bailey, I. (eds) *Turning Down the Heat: The Politics of Climate Policy in Affluent Democracies*, Palgrave, Basingstoke, pp. 104–24.

Lövbrand, Eva and Stripple, Johannes (2011) Making climate change governable: Accounting for carbon as sinks, credits and personal budgets, *Critical Policy Studies*, 5 (2): 187–200.

Lovell, Heather (2004) Framing sustainable housing as a solution to climate change, *Journal of Environmental Policy and Planning*, 6 (1): 35–56.

Lovell, Heather (2014) Climate change, markets and standards: The case of financial accounting, *Economy and Society*, 43 (2): 260–84.

Lovell, Heather (2015) *The Making of Low Carbon Economies*, Routledge, UK.

Lovell, Heather, Bulkeley, Harriet and Owens, Susan (2009) Converging agendas? Energy and climate change policies in the UK, *Environment and Planning C: Government and Policy*, 27 (1) 90–109.

Lovell, Heather and MacKenzie, Donald (2011) Accounting for carbon: The role of accounting professional organisations in governing climate change, *Antipode*, 43 (3): 704–30.

Lucas, Karen, Ross, Andrew and Fuller, Sara. (2003) *What's in a Name? Local Agenda 21, Community Planning and Neighbourhood Renewal*, YPS for Joseph Rowntree Foundation, York.

MacKenzie, Donald (2009) Making things the same: Gases, emission rights and the politics of carbon markets, *Accounting, Organizations and Society*, 34 (3–4): 440–55.

MacKenzie, Donald and Pardo-Guerra, Juan P (2014) Insurgent capitalism: Island, bricolage and the re-making of finance, *Economy and Society*, 43 (2): 153–82.

Marres, Noortje (2011) The costs of public involvement: Everyday devices of carbon accounting and the materialization of participation, *Economy and Society*, 40 (4): 510–33.

Marres, Noortje (2013) Why political ontology must be experimentalized: On ecoshowhomes as devices of participation, *Social Studies of Science*, 43 (3): 417–43.

Marres, Noortje and Lezaun, Javier (2011) Materials and devices of the public: An introduction, *Economy and Society*, 40 (4): 489–509.

Marsden, Greg and Rye, Tom (2010) The governance of transport and climate change, *Journal of Transport Geography*, 18: 669–78.

Mason, Kevin and Whitehead, Mark (2012) Transition urbanism and the contested politics of ethical place-making, *Antipode*, 44 (2): 493–516.

McGuirk, Pauline (2004) State, strategy and scale in the competitive city: A neo-Gramscian analysis of the governance of 'global Sydney', *Environment and Planning A* 36: 1019–43.

McGuirk, Pauline M. and Dowling, Robyn (2009) Neoliberal privatisation? Remapping the public and the private in Sydney's masterplanned residential estates, *Political Geography*, 28: 174–85.

McKee, Kim (2011) Sceptical, disorderly and paradoxical subjects: Problematizing the 'will to empower' in social housing governance, *Housing, Theory and Society*, 28 (1): 1–18.

Merriman, Peter (2005) Materiality, subjectification, and government: The geographies of Britain's motorway code, *Environment and Planning D: Society and Space* 23 (2): 235–50.

Miller, Peter and Rose, Nikolas (1990) Governing economic life, *Economy and Society*, 19 (1): 1–31.

Mitchell, Timothy (2008). Rethinking economy. *Geoforum*, 39 (3): 1116–21.

Mitchell, Timothy (2009) Carbon democracy, *Economy and Society*, 38 (3): 399–432.

Mol, Artur (2008) Environmental authority: Transformations and relocations in global modernity, *Desenvolvimento e Meio Ambiente*, 17: 33–46.

Murdoch, Jonathan (1997)Towards a geography of heterogeneous associations, *Progress in Human Geography*, 21 (3): 321–37.

Murdoch, Jonathan (2000) Space against time: Competing rationalities in planning for housing, *Transactions of the Institute of British Geographers*, 25 (4): 503–19.

Murdoch, Jonathan (2004) Putting discourse in its place: Planning, sustainability and the urban capacity study, *Area*, 36 (1): 50–8.

Murdoch, Jonathan and Ward, Neil (1997) Governmentality and territoriality: The statistical manufacture of Britain's 'national farm', *Political Geography*, 16 (4): 307–24.

Neal, Sarah (2013) Transition culture: Politics, localities and ruralities, *Journal of Rural Studies* 32: 60–9.

Newell, Peter and Paterson, Matthew (1998) A climate for business: Global warming, the state and capital, *Review of International Political Economy* 5 (4): 679–703.

Newell, Peter and Paterson, Matthew (2010) *Climate Capitalism: Global Warming and the Transformation of the Global Economy*, Cambridge University Press, Cambridge.

Newell, Peter, Pattberg, Philipp and Schroeder, Heike (2012) Multiactor governance and the environment, *Annual Review of Environment and Resources*, 37 (10): 365–87.

North, Peter and Longhurst, Noel (2013) Grassroots localisation? The scalar potential of and limits of the 'transition' approach to climate change and resource constraint, *Urban Studies*, 50 (7): 1423–38.

O'Brien, Karen (2012) Responding to environmental change: A new age for human geography? *Progress in Human Geography*, 35 (4): 542–9.

Okereke, Chukwumerije, Bulkeley, Harriet and Schroeder, Heike (2009) Conceptualizing climate governance beyond the international regime, *Global Environmental Politics* 9 (1): 58–78.

O'Riordan, Tim and Rowbotham, Elizabeth J. (1996) Struggling for credibility: The United Kingdom's response, in O'Riordan, T. and Jäger, J. (eds) *Politics of Climate Change: A European Perspective*, London: Routledge, pp. 228–67.

Owens, Susan (2010) Learning across levels of governance: Expert advice and the adoption of carbon dioxide emissions reduction targets in the UK, *Global Environmental Change*, 20: 394–401.

Owens, Susan and Cowell, Richard (2011) *Land and Limits: Interpreting Sustainability in the Planning Process*, 2nd edition, Routledge, London.

Painter, Joe (2006) Prosaic geographies of stateness, *Political Geography* 25: 752–74.

Painter, Joe (2010) Rethinking territory, *Antipode*, 42 (5): 1090–118.

Paterson, Matthew (1996) *Global Warming and Global Politics*, Routledge, London.

Paterson, Matthew (2001) Risky business: Insurance companies in global warming politics, *Global Environmental Politics* 1: 18–42.

Paterson, Matthew (2013) Climate re-public: Practicing public space in conditions of extreme complexity, in Best, J. and Gheciu, A. (eds) *The Public as Practice: Transformations in Global Politics*, Cambridge University Press, Cambridge.

Paterson, Matthew and Stripple, Johannes (2010) My Space: Governing individuals' carbon emissions, *Environment and Planning D: Society and Space* 28 (2): 341–62.

Paterson, Matthew and Stripple, Johannes (2012) Virtuous carbon, *Environmental Politics* 21 (4): 563–82.

Pellizzoni, Luigi (2012) In search of community: Political consumerism, governmentality and immunization, *European Journal of Social Theory*, 15 (2): 221–41.

Performance and Innovation Unit (2002) *The Energy Review*, Performance and Innovation Unit, London.

Prince of Wales (2014) UK CLG: University of Cambridge Programme for Sustainability Leadership. Available online: www.princeofwales.gov.uk/the-prince-of-wales/the-princes-charities/university-of-cambridge-programme-sustainability/case-study/uk-clg (accessed February 2015).

Repowering London (2014) Roupell Park Brixton Hill. Available online: www.repowering.org.uk/projects/roupell-park#sthash.5XbTUNSj.dpuf (accessed February 2015).

Rice, Jennifer L (2014) An urban political ecology of climate change governance, *Geography Compass*, 8 (6): 381–94.

River Cottage (2013) Energyshare. Accessed online: www.rivercottage.net/about/campaigns/energyshare/ (accessed March 2013, no longer available online).

Rose, Nikolas (1999) *Powers of Freedom: Reframing Political Thought*, Cambridge University Press, Cambridge.

Rose, Nikolas and Miller, Peter (1992) Political power beyond the state: Problematics of government, *British Journal of Sociology*, 43 (2): 173–205.

Rose-Redwood, Reuben (2012) With numbers in place: Security, territory, and the production of calculable space, *Annals of the Association of American Geographers*, 102 (2): 295–319.

Rosenau, James (2000) Change, complexity and governance in a globalizing space, in Pierre, Jon (ed.) *Debating Governance*, Oxford University Press, Oxford.

Rosenau, James N. (2002) NGOs and fragmented authority in globalizing space, in Ferguson, Y. H. and Barry Jones, R. J. (eds), *Political Space: Frontiers of Change and Governance in a Globalizing World*, State University of New York Press, Albany, pp. 261–80.

Rubio, Fernando D and Fogue, Uriel (2013) Technifying public space and publicizing infrastructures: Exploring new urban political ecologies through the Square of General Vara del Rey, *International Journal of Urban and Regional Research*, 37 (3): 1035–52.

Ruggie, John (2004) Reconstituting the global public domain: Issues, actors and practices, *Journal of International Relations* 10 (4): 499–531.

Rutherford, Stephanie (2007) Green governmentality: Insights and opportunities in the study of nature's rule, *Progress in Human Geography*, 31 (3): 291–307.

Rutland, Ted and Aylett, Alex (2008) The work of policy: Actor networks, governmentality, and local action on climate change in Portland, Oregon, *Environment and Planning D: Society and Space*, 26: 627–46.

Rutter, Jill, Marshall, Edward and Sims, Sam (2012) The 'S' factors: Lessons from IFG's policy success reunions, Institute for Government. Available online: www.instituteforgovernment.org.uk/sites/default/files/climate_change_act.pdf (accessed February 2015).

Rydin, Yvonne (2007) Indicators as a governmental technology? The lessons of community-based sustainability indicator projects, *Environment and Planning D: Society and Space*, 25: 610–24.

Saunders, Clare (2012) Reformism and radicalism in the Climate Camp in Britain: Benign coexistence, tensions and prospects for bridging, *Environmental Politics*, 21 (5): 829–46.

Scott, James C. (1998) *Seeing Like a State: How Certain Schemes to Improve the Human Condition Have Failed*, Yale University Press, New Haven and London.

Sending, Ole J. and Neumann, Iver B (2006) Governance to governmentality: Analyzing NGOs, states, and power, *International Studies Quarterly* 50: 651–72.

Seyfang, Gill, Park, Jung Jin and Smith, Adrian (2013) A thousand flowers blooming? An examination of community energy in the UK, *Energy Policy* 61: 977–89.

Shove, Elizabeth (2010) Beyond the ABC: Climate change policy and theories of social change, *Environment and Planning A*, 42 (6): 1273–85.

Smith, Blair (2013) HSBC Climate Champion Program. Available online: www.youtube.com/watch?v=dXgqx-q9ZB0, Minutes 2.15-3 (accessed May 2015).

Smith, Timothy M. and Fischlein, Miriam (2010) Rival private governance networks: Competing to define the rules of sustainability performance, *Global Environmental Change* 20 (3): 511–22.

Soil Association (2014) Low carbon farming. Available online: www.soilassociation.org/lowcarbon (accessed February 2015).

Stern, Nicholas. (2007) *The Economics of Climate Change: The Stern Review*, Cambridge University Press, Cambridge.

Stripple, Johannes (2007) The stuff of international relations? Process philosophy as meta-theoretical reflection on security, territory and authority. Sixth Pan-European International Relations Conference. Turin, Italy. Available online: www.eisa-net.org/be-bruga/eisa/files/events/turin/Stripple-Stripple%20Stuff%20of%20IR.pdf (accessed February 2015).

Stripple, Johannes and Bulkeley, Harriet (eds) (2014) *Governing the Climate: New Approaches to Rationality, Power and Politics*, Cambridge University Press, New York.

Summerville, Jennifer A., Adkins, Barbara A. and Kendall, Gavin (2008) Community participation, rights, and responsibilities: The governmentality of sustainable development policy in Australia, *Environment and Planning C: Government and Policy* 26 (4): 696–711.

Swyngedouw, Eric (2010) Apocalypse forever? Post-political populism and the spectre of climate change, *Theory, Culture and Society* 27 (2–3): 213–32.

Telegraph (2014) James Dyson suggests leaving the EU over vacuum cleaners, 5 September. Available online: www.telegraph.co.uk/technology/news/11075490/James-Dyson-suggests-leaving-the-EU-over-vacuum-cleaners.html (accessed February 2015).

Telier, A. (2012) Drawing things together, *Interactions* 19 (2): 34–7.

Tesco (2009a) Building environmental stores: Cheetham Hill, Manchester 2009. Accessed online: www.tescoplc.com/plc/storage/pdf/build_env.pdf (accessed March 2012, no longer available online).

Tesco (2009b) Building environmental stores: Zero-carbon Ramsey, Cambridgeshire 2009. Accessed online: www.tesco.com/assets/greenerliving/content/documents/pdfs/Environmental%20A6Leaflet.pdf (accessed March 2012, no longer available online).

Tesco (2010) *Corporate Responsibility Report 2010*, Tesco PLC, Cheshunt, Hertfordshire.

Tesco (2012a) *Corporate Responsibility Review 2012*, Tesco PLC, Cheshunt, Hertfordshire.

Tesco (2012b) *Product Carbon Footprint Summary*. Available online: www.tescoplc.com/ assets/files/cms/.Tesco_Product_Carbon_Footprints_Summary%281%29.pdf (accessed February 2015).

Tesco (2014) Reducing our impact on the environment. Available online: www.tescoplc.com/index.asp?pageid=634#tabnav (accessed February 2015).

Timmermans, Stefan and Epstein, Steven (2010) A world of standards but not a standard world: Toward a sociology of standards and standardization, *Annual Review Sociology*, 36: 69–89.

Thornley, Andy, Rydin, Yvonne, Scanlon, Kath and West, Karen (2005) Business privilege and the strategic planning agenda of the Greater London Authority, *Urban Studies*, 42 (11): 1947–68.

Tyne Rivers Trust (2012) Hydro project must not destroy river's assets. Letter to the *Hexham Courant*, 19 December. Available online: www.hexhamcourant.co.uk/2.2993/letters/hydro-project-must-not-destroy-river-s-assets-1.907877 (accessed February 2015).

Uitermark, Justus (2005) The genesis and evolution of urban policy: A confrontation of regulationist and governmentality approaches, *Political Geography*, 24: 137–63.

UNEP [United Nations Environmental Programme] (2014) UNEP Finance Initiative: About. Available online: www.unepfi.org/about/background/ (accessed February 2015).

Walker, Gordon P. and Devine Wright, Patrick (2008) Community renewable energy: What does it mean? *Energy Policy* 36: 498–99.

Walker, Gordon P., Hunter, Sue, Devine-Wright, Patrick, Evans, Bob and Fay, Helen (2007) Harnessing community energies: Explaining and evaluating community-based localism in renewable energy policy in the UK, *Global Environmental Politics*, 7 (2): 64–82.

Walters, William (2012) *Governmentality: Critical Encounters*, New York: Routledge.

Whatmore, Sarah (2013) Earthly powers and affective environments: An ontological politics of flood risk, *Theory, Culture & Society*, 30 (7/8): 33–50.

Whitehead, Mark (2009) *State, Science and the Skies: Governmentalities of the British Atmosphere*, RGS-IBG Book Series, Wiley-Blackwell, Chichester, UK.

Index

Lightning Source UK Ltd.
Milton Keynes UK
UKOW04n0747290617
304327UK00015B/340/P

9 781107 038653